PREFACE

When teachers communicate an enthusiasm about numbers and mathematics to their students, wonderful learning can happen. Where teachers recognize that understanding of numbers comes gradually, students develop an understanding of how numbers, objects, counting and computation are related.

All too often, however, first and second grade teachers feel compelled to teach to the test rather than to teach in ways that meet the developmental needs of their students. Then enthusiasm for numbers and curiosity about how numbers work is lost among the numerical representations that do not connect with the ways in which young learners think about numbers.

Playing with numbers is an important experience if students are to develop number sense. Then, talking about the internal language of number play becomes an important learning tool, as students generalize about their experiences. In this unit, we encourage students to manipulate and organize objects. Then we encourage students to talk and write about numbers in their own ways, using words and pictures. We encourage invented notation to record their thinking and their findings; and we model standard notation. With second graders, we compare invented notation with standard notation and talk about the advantages of each.

Young learners seem to benefit from using many different visual models as they develop a sense of "how many" and "how many more." In this unit, we support the organized exploration of numbers by introducing tools that help students organize objects into visual arrays: object graphs and ten-frames. Students see the utility of these organizational tools only when they recognize that counting more than a few of an object can be cumbersome.

Some students quickly understand how to make use of graphs and ten-frames in order to add and compare quantities. Others need more time to understand how these tools can be helpful. In either case, do not give in to the temptation to move students into standard notation prematurely.

Children come to understand number meanings gradually. To encourage these understandings, teachers can offer classroom experiences in which students first manipulate physical objects and then use their own words to explain their thinking. . .

Emphasizing exploratory experiences with numbers that capitalize on the natural insights of children enhances their sense of mathematical competency, enables them to build and extend number relationships, and helps them to develop a link between their world and the world of mathematics.

Children's experiences with numbers are most beneficial when the numbers have meaning for them.

NCTM Standards
page 38

I have noticed that young learners do not "progress" in a linear fashion from physical to visual to symbolic representations for numbers. Instead, they seem to move from one mode to another and back again. Having observed the value of simultaneous representations, I encourage all students to show me a picture, whether for a story or for an equation. In this way, we keep alive the idea that numbers can be represented in different ways and that a picture of a number often reveals something important about that number that might be helpful in another context.

Play continues to be an important learning vehicle for primary grade students. Without the play experiences, students do not develop a basis for understanding numbers. When we create experiences that rely on playing with numbers, when we encourage visual and physical representations for numbers, when we encourage students to model how numbers of objects combine and compare, then we are helping students develop a foundation for understanding computation.

Addition and Subtraction

LEVEL A

by Christine Losq

A Houghton Mifflin Company
Wilmington, Massachusetts

Acknowledgments: The development of these materials would not have been possible without the inspiration of the teachers and students who have shared their experiences and their insights. In the classrooms that used these materials, students were engaged in developing meaningful problem-solving strategies. Students and teachers were enthusiastic about math. Students were encouraged to give voice to their curiosity; they exhibited the habits of responsibility and cooperation that make for a true community of learners.

My heartfelt thanks to those who have used, critiqued, and helped shape these activities—especially Patrick Lee, Karen McLeod, Kim Robertson, Avery Walker, May Yee, Martha Wood and their students.

I am particularly grateful to the many students from whom I continue to learn—the first and second grade students taught by May Yee, Martha Wood and Kim Robertson at Haskell Elementary School and Haskell Magnet School, Granada Hills, California, who shared their wonderful ideas about adding, subtracting, and counting fish; Karen McLeod's students at Harding Avenue Elementary School, Blacksburg, Virginia, who generously shared their insights, their questions, their strategies, and their writing; Avery Walker's students at Ohlone Elementary School, Palo Alto, California, who helped shape the money lessons with their remarkable business sense; and Patrick Lee's students, also at Ohlone, who enthusiastically shaped Swimmy Math and Marshmallow Math and worked so hard to share their ideas with me on paper.

To all of these members of my learning circle I dedicate this book.

Grateful acknowledgment is made for permission to reprint original or copyrighted material:
Ten-frame Math lessons, **Marshmallow Math**, and accompanying reproducible student pages reprinted with permission from ***Ten-frame Math*** ©CSL Associates, Inc. All rights reserved.

Cover: Photo by David S. Waitz. Design and production by Bill SMITH STUDIO: Brian Kobberger, Art Director; Sandra Schmitt, Designer; Justine Price, Photo Editor.

Text: Design and production by Flying Pages, Inc.

Illustrations copyright ©Nicolas Losq. Used with permission of the artist.

Printed in the United States of America

International Standard Book Number: 0-669-44444-8

1 2 3 4 5 6 7 8 9 0 PO 02 01 00 99 98 97

URL address: http://www.greatsource.com/

TABLE OF CONTENTS

WELCOME TO MATHZONES

Welcome to MathZones! When you use this book, you are joining the many teachers who are setting aside their standard textbooks and using activity-based learning to bring addition and subtraction to life. The activities in this book will help you create an exciting and supportive learning environment with your students.

Imagine your classroom. Students are on task, sharing ideas and strategies, checking results to see if they make sense. Imagine having time to work with individuals and small groups of students. Imagine your feeling of success when your students demonstrate understanding of multi-digit addition and subtraction strategies—composing and decomposing tens, picturing equations, developing number sense.

As your students are actively involved in each lesson, you will be able to observe learning in progress. You will be able to observe as your students develop the essential problem-solving strategies that reveal the thinker in every student.

Using these activities will help your students to:

- sort, classify, and count
- explain problem situations
- understand data
- visualize and organize data
- connect addition and subtraction ideas to everyday situations
- connect math with other curriculum areas

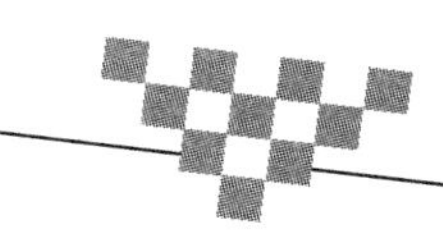

When we, as teachers, pull our chairs alongside [our students] and try to understand their ways of understanding, when we search for the logic of their errors and the patterns of their growth, then we no longer spin our wheels.

—*Lucy Calkins,* ***The Art of Teaching Writing*** *(Heinemann: Portsmouth, NH)*

USING MATHZONES

USING THIS UNIT

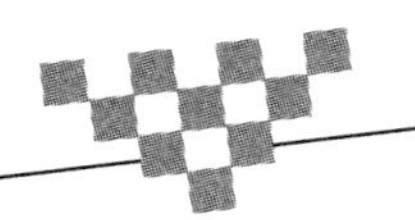

Four kinds of learning experiences are included in this unit:

- ***Assessments***
- ***Whole-group activities***
- ***Cooperative-group activities***
- ***Literature-based lessons***

Addition and Subtraction contains lessons that develop addition and subtraction concepts in visual and story settings. The lessons are designed to help students develop an understanding of addition and subtraction.

Assessments are open-ended and provide ongoing feedback about students' evolving understanding of addition and subtraction. The assessments in this unit are as follows:

Assessment 1: Write About Addition and Subtraction
Students share prior knowledge and venture some generalizations about addition and subtraction.

Assessment 2: Write and Solve Fish Stories
Students share their understanding of addition and subtraction.

Assessment 3: What Have We Learned?
Students demonstrate what they have learned about addition and subtraction.

For more information about **Assessment,** see page 17.

A **whole-group** learning activity begins most lessons. Whole-group instruction is usually followed by small, heterogeneous groups of students working together on another activity. **Cooperative learning** or small-group instruction adds an exciting dimension to the learning of mathematics. Student benefits include:

- face-to-face interaction.
- group processing.
- individual accountability.
- increased self-esteem.
- development of social skills.
- higher-level thinking.
- increased individual achievement.

Literature-based lessons in this unit suggest these books: ***Swimmy*** by Leo Lionni and ***Arthur's Funny Money*** by Lillian Hoban. Additional titles relating to the topic are provided in the **Bibliography** on page 125.

In addition, this unit teaches students many useful skills and applications of addition and subtraction.

Students:

- skip count.
- use a hundred chart.
- find patterns and organize subtraction facts.
- practice addition and subtraction.
- explore adding two-digit numbers.
- group and arrange objects.
- use ten-frames.
- count and classify shapes.
- create a graph.
- solve story problems.
- write addition equations.
- add and subtract money.

Unit Pacing

Addition and Subtraction works well as a four-to-six week unit.

To help you plan the unit for your class, a **Skills Matrix** chart is provided on pages 6 and 7.

This unit suggests a sequence of teaching. However, we strongly suggest that you evaluate your students' needs and adjust the pacing of this unit and the selection of lessons to them.

USING THE LESSON PLAN

Lesson plans include:

- ***Materials***
- ***Advance Preparation***
- ***Objectives***
- ***Getting Started***
- ***Wrapping Up***
- ***Write About Math***
- ***Teaching Tips***
- ***Extensions/ Homework***
- ***What Really Happened***
- ***Sharing Ideas and Strategies***

1 Each lesson is designed for easy planning and implementation. The first page of each lesson shows a list of **Materials.** It is conveniently placed in the margin for easy reference and lesson management. This list includes supplies that are required for both teachers and students.

2 **Advance Preparation** suggests ways teachers could prepare for the lesson and avoid any delays when introducing the lesson.

3 The second page of each lesson begins with a list of **Objectives**, clearly-written statements of what students will learn.

4 **Getting Started** launches the lesson. Ideas for warming up for the lesson are presented. Questions may be included to help students think critically and become independent learners. Guidelines for coaching students are often provided.

5 **Wrapping Up** offers ideas to close out a lesson. Sometimes a writing prompt suggestion is described under **Write About Math**. Students are encouraged to write in their math journals throughout the unit.

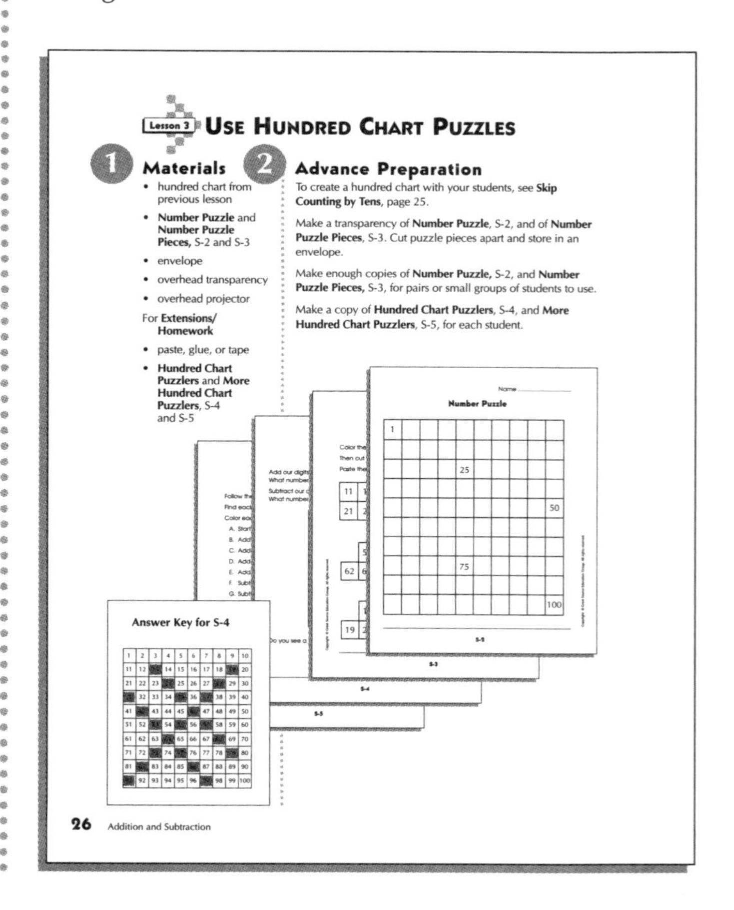
Lesson 3 USE HUNDRED CHART PUZZLES

1 **Materials**

- hundred chart from previous lesson
- **Number Puzzle** and **Number Puzzle Pieces,** S-2 and S-3
- envelope
- overhead transparency
- overhead projector

For **Extensions/ Homework**

- paste, glue, or tape
- **Hundred Chart Puzzlers** and **More Hundred Chart Puzzlers**, S-4 and S-5

2 **Advance Preparation**

To create a hundred chart with your students, see **Skip Counting by Tens**, page 25.

Make a transparency of **Number Puzzle**, S-2, and of **Number Puzzle Pieces**, S-3. Cut puzzle pieces apart and store in an envelope.

Make enough copies of **Number Puzzle,** S-2, and **Number Puzzle Pieces,** S-3, for pairs or small groups of students to use.

Make a copy of **Hundred Chart Puzzlers**, S-4, and **More Hundred Chart Puzzlers**, S-5, for each student.

26 Addition and Subtraction

❻ **Teaching Tips** offer advice on classroom management, questions to facilitate discussion, extension activities, or an informal assessment idea.

❼ **Extensions/Homework** activities are provided at the end of some lessons. In addition, the feature **Extensions and Homework** may appear at the end of a section. They relate to all lessons in the section.

❽ **What Really Happened** provides vignettes from the classroom, sharing ways in which students transformed activities to meet their learning needs. These glimpses into one particular classroom are not intended to be prescriptive, but are provided as a way of sharing teaching and learning experiences.

❾ **Sharing Ideas and Strategies** offer real-life student work—and comments that help interpret it as well as ideas for assessment.

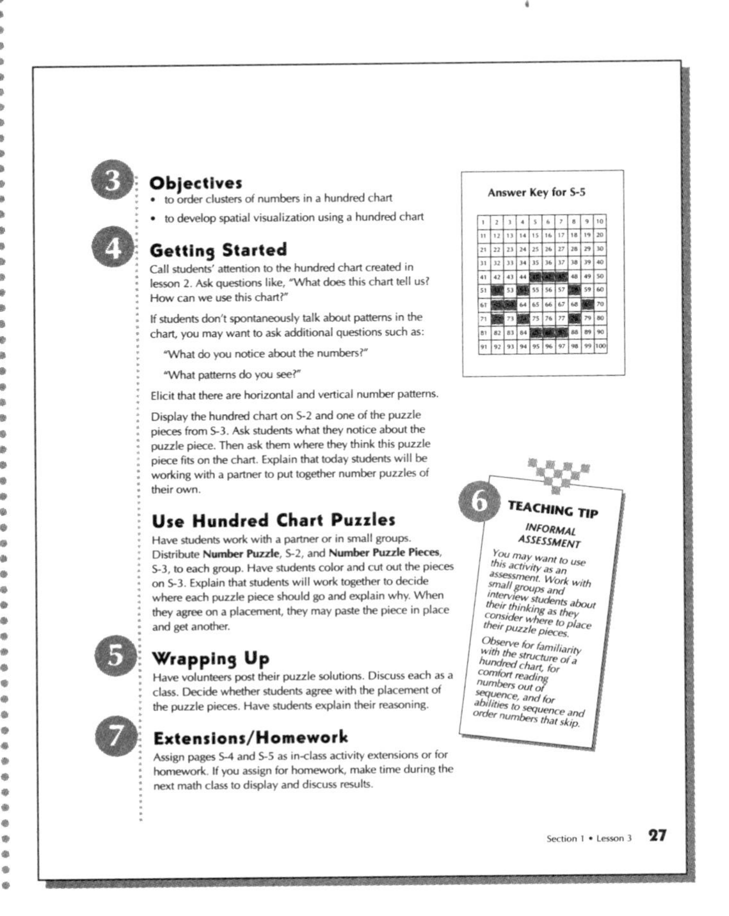

❸ **Objectives**

- to order clusters of numbers in a hundred chart
- to develop spatial visualization using a hundred chart

❹ **Getting Started**

Call students' attention to the hundred chart created in lesson 2. Ask questions like, "What does this chart tell us? How can we use this chart?"

If students don't spontaneously talk about patterns in the chart, you may want to ask additional questions such as:

"What do you notice about the numbers?"

"What patterns do you see?"

Elicit that there are horizontal and vertical number patterns.

Display the hundred chart on S-2 and one of the puzzle pieces from S-3. Ask students what they notice about the puzzle piece. Then ask them where they think this puzzle piece fits on the chart. Explain that today students will be working with a partner to put together number puzzles of their own.

Use Hundred Chart Puzzles

Have students work with a partner or in small groups. Distribute **Number Puzzle**, S-2, and **Number Puzzle Pieces**, S-3, to each group. Have students color and cut out the pieces on S-3. Explain that students will work together to decide where each puzzle piece should go and explain why. When they agree on a placement, they may paste the piece in place and get another.

❺ **Wrapping Up**

Have volunteers post their puzzle solutions. Discuss each as a class. Decide whether students agree with the placement of the puzzle pieces. Have students explain their reasoning.

❼ **Extensions/Homework**

Assign pages S-4 and S-5 as in-class activity extensions or for homework. If you assign for homework, make time during the next math class to display and discuss results.

Answer Key for S-5

1	2	3	4	5	6	7	8	9	10
11	12	13	14	15	16	17	18	19	20
21	22	23	24	25	26	27	28	29	30
31	32	33	34	35	36	37	38	39	40
41	42	43	44				48	49	50
51		53		55	56	57		59	60
61			64	65	66	67	68		70
71		73		75	76	77		79	80
81	82	83	84				88	89	90
91	92	93	94	95	96	97	98	99	100

❻ **TEACHING TIP**

INFORMAL ASSESSMENT

You may want to use this activity as an assessment. Work with small groups and interview students about their thinking as they consider where to place their puzzle pieces.

Observe for familiarity with the structure of a hundred chart, for comfort reading numbers out of sequence, and for abilities to sequence and order numbers that skip.

Section 1 • Lesson 3 **27**

SKILLS MATRIX

Concepts and Skills / Lesson	1. Write About Addition and Subtraction—Assessment 1	2. Skip Counting by Tens	3. Use Hundred Chart Puzzles	4. Keep the Difference	5. Organize and Graph Data	6. Introduce Ten-frame Math	7. Group and Count Objects	8. Estimate Fish Area	9. Make a New Ten	10. Make a Fish Collage
Write about math	✔	✔				✔			✔	✔
Problem solving	✔	✔	✔	✔	✔	✔	✔	✔	✔	✔
Communication	✔	✔	✔	✔	✔	✔	✔	✔	✔	✔
Reasoning	✔	✔	✔	✔	✔	✔	✔	✔	✔	✔
Sorting and classifying					✔					
Counting		✔	✔	✔		✔	✔	✔		
Skip counting		✔						✔		
Number patterns and number sense	✔	✔	✔		✔	✔			✔	
Collecting and organizing data					✔	✔	✔	✔	✔	
Composing and decomposing tens				✔	✔	✔	✔		✔	✔
Addition facts				✔		✔				✔
Adding three addends										
2-digit addition	✔					✔	✔		✔	✔
Adding with money										
Graphing					✔					
Estimating						✔	✔	✔		
Subtraction facts	✔			✔						
2-digit subtraction	✔									
Subtracting with money										
Mental math						✔			✔	
Consumer math										

11. Count, Add, and Organize Collage Shapes	12. Graph Collage Shapes	13. Write and Solve Fish Stories—Assessment 2	14. Record Marshmallows and Macaroni	15. Ten-frame Addition	16. Marshmallow Take-Away	17. Review Money Equivalents	18. Count Arthur's Money	19. Open for Business	20. Keep Track of Arthur's Money	21. Start a Business	22. What Have We Learned—Assessment 3
✔	✔	✔				✔	✔	✔	✔	✔	✔
✔	✔	✔	✔	✔	✔	✔	✔	✔	✔	✔	✔
✔	✔	✔	✔	✔	✔	✔	✔	✔	✔	✔	✔
✔	✔	✔	✔	✔	✔	✔	✔	✔	✔	✔	✔
✔						✔	✔	✔			✔
✔							✔	✔	✔		
✔	✔	✔	✔		✔						✔
✔								✔	✔		✔
	✔	✔		✔	✔		✔		✔		✔
✔				✔			✔				✔
	✔			✔			✔				✔
✔	✔	✔	✔	✔	✔	✔	✔	✔	✔	✔	✔
						✔	✔	✔	✔	✔	
	✔										
					✔					✔	✔
					✔					✔	✔
						✔	✔	✔	✔	✔	
	✔			✔	✔		✔	✔	✔	✔	
						✔	✔	✔	✔	✔	

CLASSROOM ORGANIZATION

This unit was used in a variety of classrooms. As a general rule, math time was scheduled four days each week for a minimum of forty-five minutes each day and up to eighty minutes (the time between morning recess and lunch or the afternoon after lunch until afternoon recess). Students could choose to work on other assignments when the math activity had been completed. Because the writing assignments coincided with language arts objectives, teachers were able to integrate teaching objectives without having to make "extra time" for math.

The philosophy in the classroom emphasized that we all learn from one another. Therefore, once an activity had been launched, students were encouraged to work in pairs or triads or, in some cases, in small groups directed by an adult. As activities came to closure, each pair or triad presented their solution strategies to the rest of the group. Classmates were encouraged to ask questions, compare strategies, and reconcile solutions.

Students were frequently asked to record their thinking about math. In first grade groups, recording ideas was often facilitated by an adult or older student who might take dictation.

MATERIALS

The activities in this unit emphasize hands-on learning. Most lessons refer to student activity pages, which are indicated by the prefix "S." For your convenience, we have provided reproducible **Student Activity Pages** at the back of the book. (See page 129 for an overview.) You will also find reproducible copymasters in the **Teacher Resources** section. These include centimeter graph paper, and ten-frame grids, and more. See page 107 for an overview.

Advance Preparation ideas and reminders are provided on the **Materials** page for each lesson and often include reproducible student activity pages. In several lessons, you need to make a model for classroom display.

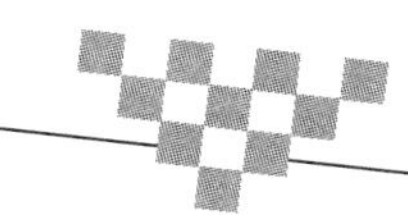

TEACHING TIP

INVOLVING FAMILIES

If your classroom budget does not allow for purchase of materials, ask local merchants or service organizations if they would be willing to join in this activity and make donations of materials.

As an alternative, ask families to contribute toward the materials by bringing in one of the suggested items.

Classroom Materials

Classroom materials that you will need for this unit include:

- paper pattern block shapes (use colored card stock)
- a roll of sentence strip tagboard
- number cubes (standard 6-sided and blank 6-sided)
- transparency film
- scissors (a pair for each student)
- index cards or card stock
- chart paper
- paper cups, paper plates
- small paper lunch bags
- crayons or markers
- glue or paste
- goldfish crackers
- miniature marshmallows
- elbow macaroni
- countable objects (counters, centimeter cubes, pennies)
- play coins or coin stamps
- calculators (optional)

Recommended Literature Library

Swimmy by Leo Lionni, Dragonfly Books™, Alfred Knopf, NY

Arthur's Funny Money by Lillian Hoban, Harper Trophy, NY

For more books about addition and subtraction, refer to the Bibliography starting on page 125.

THINKING AND TALKING MATH

Opportunities to talk and write about mathematics create a powerful teaching and learning tool. Throughout this unit, students are asked to explain, both orally and in writing, their ideas about patterns, counting, adding and subtracting. Students are always asked to explain their solution strategies. We are always asking questions like:

1. What are you trying to find out? Restate the problem in your own words.
2. What do these numbers say? What does this picture say? Show me what you are thinking.
3. What strategy did you use? Describe your method so someone else could try it.
4. How do you know your answer is right? Could there be more than one right answer? How do you know?
5. Did anyone solve this problem in a different way? How many different ways can we find to show this problem?

When these questions are addressed regularly, students come to understand that thinking about their strategies and reflecting on the logic of an answer are more important than simply finding a numerical solution. They come to understand that a problem can be represented many different ways. They begin to see logical connections between verbal language and numbers.

Explaining their thinking can be a challenge for first and second-graders, especially if they are used to one right answer in arithmetic. "I just know it!" is a common response from children with good rote recall. "I don't know. Is it wrong?" is a typical response from children who do not trust their mathematical thinking. All children benefit from experiences that model the value of talking about numbers.

Writing about mathematics means responding to questions like, "What is addition all about?" "When would we ever need to subtract?" "Describe a pattern that you see." "What surprised you about your answer?" "What did you learn from this activity?"

Writing about math also means recording information in different formats—graphs, tallies, equations, pictures, stories.

Finally, writing about math means linking different representations for number ideas. Understanding that a picture can be described in words, and that a picture can be described in numbers is a leap for first and secondgraders that can be nurtured best through language.

FOUR-STEP WRITING PLAN

The four-step writing plan outlined below is a valuable learning tool that helps students develop their own problem-solving strategies. You can use these questions as discussion prompts or as writing prompts. This plan also provides questioning strategies that help you coach students to become powerful, independent problem solvers.

1. *What are you trying to find out? Restate the problem in your own words.*

When students restate a problem, they are better able to understand what the problem is about. Restating a problem encourages students to make it their own. Restating a problem also encourages students to ask questions they may have about the situation and/or about the data.

2. *What are your first impressions? How do you think you might solve this problem?*

Encouraging students to express their first impressions allows them to consider how this problem is like other problems they have encountered. Some students need to write about their anxieties before they can gain the courage to take on a challenge.

3. *How did you find your answer? Describe your strategy so that someone else can try it.*

When students explain their strategies, they reflect on their thinking. Here you can assess for conceptual understandings and encourage students to develop a repertoire of problem-solving strategies that is meaningful to the student. Most importantly, when students explain their strategies, they come to value their own thinking.

4. *How do you know your answer is right? Could there be more than one right answer? Why or why not?*

Encouraging students to examine their answers helps develop independent thinking. When students explain the reasonableness of their answers, they validate their methods and computational accuracy. As students explain how they know their answer is right, they learn to rely on themselves rather than on the approval of the teacher. When students discuss answers, they learn to develop listening skills; they build their own language skills; they learn to evaluate their own thinking.

USING MATH JOURNALS

Math journals help students keep a record of their thinking and of the evolution of their ideas. For each Write About Math prompt, we ask students to start on a clean page and clearly label the date and the topic.

We have tried various kinds of notebooks and find that steno pads work well. The pages are large enough to encourage students to "think in writing," yet are not so large that much paper is wasted. Steno pads are also relatively inexpensive and easy to find in bulk in office supply stores.

We have also had good experiences with spiral-bound graph paper notebooks. Graph paper helps organize writing and sketches. Equations can be aligned along a vertical line. Tables and charts can be easily set up on a grid. However, graph paper journals are relatively expensive.

When you consider what you will have your students use for math journals, you may want to keep in mind:

Size: A math journal need not have more than 32 pages. If your students are using student activity pages, you may prefer that they purchase a folder or spiral notebook to keep all related papers together in one place.

Cost: Will students supply their own journals or does your school budget allow for this purchase? Steno pads are relatively inexpensive.

Creative Alternatives: Does your school's supply room provide blue composition books? If these are available, have holes drilled so students can keep them in their binders.

Does your supply shelf contain construction paper, plain paper, or newsprint? If so, have students make their own 32-page journals by stapling or fastening together sheets of paper inserted into a sheet of construction paper that has been folded in half.

Whatever format you choose, encourage students to use their journals. Many of our lessons have a writing prompt or a problem for student journals. Journals can be helpful in showing student growth as the year progresses. If you use portfolio assessment, a journal can be an important component.

TEACHING AN INTEGRATED CURRICULUM

Many teachers are embracing the idea of integrating literature, language arts, science, math, and social studies. The literature selected for this unit work well with many of the lessons. Refer to the Bibliography on page 125 for more ideas.

Integrate **science** with a unit about animals. Students can learn about animal families.

It's easy to integrate **language arts** with addition and subtraction. Invite students to write addition and subtraction stories about things, people and animals.

You can also integrate **art** by having students illustrate addition and subtraction stories.

INTEGRATING MATHEMATICS INTO THE CURRICULUM

Addition and Subtraction provides addition and subtraction lessons that are connected to the real world.

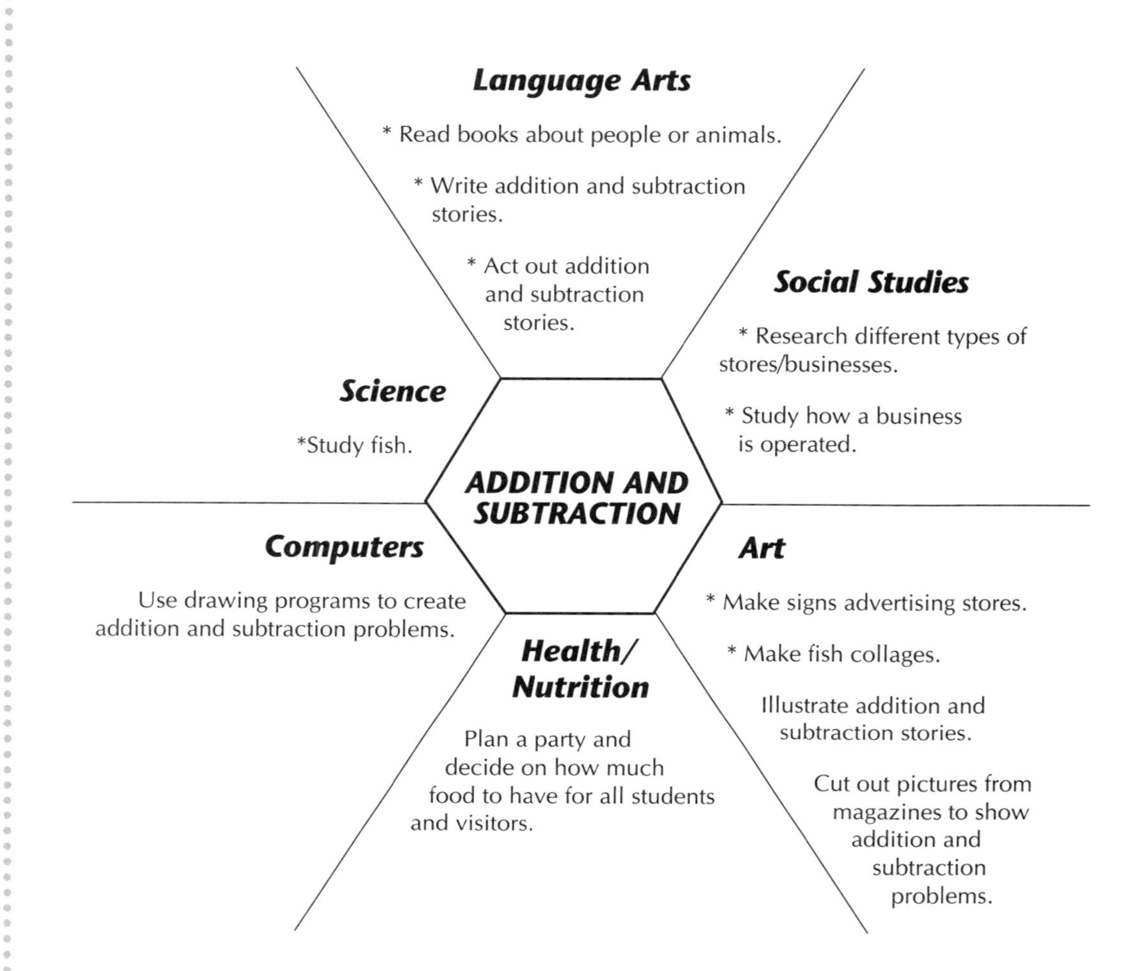

* Integrations included in this unit.

TEACHER TALK

Manipulatives and visual models are essential tools for teaching mathematics at any grade level. For the activities in this unit, you will want to have countable objects available for each student. Because commercially available manipulatives are expensive and often are not available in sufficient quantities in every classroom, we have built our activities around readily available "consumable" manipulatives that engaged our students' imaginations—fish-shaped crackers, marshmallows, and macaroni.

Use tenframes.
Count your marshmallows.
Record what you see.
Write the number.
10

Use tenframes.
Count your macaroni.
Record what you see.
Write the number.
13

How many marshmallows and macaronis do you have in all?
Write an equation in the box.

$$\begin{array}{r} 10 \\ +13 \\ \hline 23 \end{array}$$

Ten-frames and number lines are some of the visual models that help children think about organizing objects for counting. Visual models can be created with manipulatives and provide a natural bridge to pictorial models. Number lines will probably be a familiar tool for your students. Ten-frames may be new.

The ten-frame is a wonderful visual tool for seeing how we compose and decompose tens when we add and subtract. When you introduce **Ten-frame Math** (see pages 38–39), be sure to give students an opportunity to examine and describe the tool. Encourage students to talk about how the "windows" in a ten-frame are arranged and how ten-frames might be useful when we add or subtract or compare quantities.

Whenever you model a mathematical concept with manipulatives and other tools, plan up to a full class period for students to examine, explore, and discuss the materials. For example, even though your students have already used pattern blocks before, encourage conversation about how to use pattern blocks for **Make a Fish Collage,** page 54–55. When we give students time to examine concrete materials, we are encouraging their desire to rediscover and learn. When we give them time to "play," we are encouraging them to use their observation skills and creativity. When we ask them to describe and classify their observations, we are helping them value the skills that make for powerful thinkers.

TEACHING TIP

USING MANIPULATIVES MEANINGFULLY

A surprising number of very young students hesitate to use manipulatives to explore arithmetic ideas, even when they might want to.

You can overcome this prejudice by encouraging all students to "play" with materials and share their observations with a partner.

Sometimes it is useful to follow up on discovery time with a brief group discussion. Encourage close observation and discussion with questions like, "What do you notice?"

Make a list of students' observations and post for reference.

FAMILY INVOLVEMENT

Today we are promoting teaching mathematics in a way substantially different from the ways in which most of us were taught. The new pedagogy emphasizes problem solving over rote procedures, explanations of thought processes in addition to mere answers, and teamwork in the learning process rather than competition. The pedagogy endorsed by the NCTM *Standards* and advocates of constructed learning seeks to develop critical thinking as a habit of mind for all students.

Building parent support for this way of teaching begins with acknowledging parents' concerns about the new math instruction that seems foreign to their experience and their expectations. It is helpful when addressing parent concerns to begin by explaining the purposes of this unit, emphasizing the importance of listening to children's logic and encouraging children to think for themselves.

This addition and subtraction unit offers family members many opportunities to get involved. If you are fortunate to have volunteers in the classroom, ask them to prepare materials for the lessons and help children cut and write as needed.

However, this unit also works fine if families cannot be involved. Most lessons are set up so that students progress by doing hands-on lessons. They also help each other learn.

Extensions and homework suggestions can be found throughout the unit.

ASSESSMENT

Assessment is available throughout this unit and is, for the most part, ongoing and informal. Pretests, like **Write About Addition and Subtraction,** invite students to think broadly about the topic and give teachers feedback about the class' starting point.

Informal, ongoing assessments that include talking and writing about math challenge students to think about their methods and the reasonableness of their answers. For example, Assessment 2, **Write and Solve Fish Stories,** evaluates students' understanding of addition and subtraction. Writing their own story problems allows students to choose the magnitude of number that is comfortable for them. All of these elements together give us a fuller profile of where students are developmentally, in their evolving sense of numbers and number relationships, and in their confidence and willingness to take risks with larger numbers. In Assessment 3, **What Have We Learned?,** students critique and revise their earlier statements about addition and subtraction to incorporate new learning.

The act of teaching should be founded on dialogues between teachers and students... Assessment must be more than testing; it must be a continuous, dynamic, and often informal process.

NCTM Standards
page 203

Many of the written and graphic assignments will give you insights into student progress. If you are required to issue grades, the sum of these projects can provide a basis for grading. For group or collaborative projects, you might also assign a grade to each group member for participation.

Whatever assessment tools you choose, be assured that you will measure much progress as your students learn about addition and subtraction.

EXPLORING ADDITION AND SUBTRACTION

Assessment 1

Lesson 1

Write About Addition and Subtraction

Materials

- **Write About Addition and Subtraction**, S-1
- math journals
- pencils
- tagboard
- marker

Advance Preparation

Make a copy of **Write About Addition and Subtraction**, S-1, for each student.

Name ____________ Date ____________

Write About Addition and Subtraction

What do you know about adding?
Write about a time when you had to add.

____
____
____
____
____
____

What do you know about subtracting?
Write about a time when you had to subtract.

____
____
____
____
____
____
____

S-1

Objectives

- to assess students' prior knowledge of addition and subtraction
- to share ideas about a math topic

Getting Started

Begin the unit by having students think about what they already know about adding and subtracting. Have students take a few minutes to note their ideas in their journals. Circulate and coach individual students as needed.

Write About Addition and Subtraction

After students have had a chance to jot down their notes, ask them to share their ideas in a class discussion. Make a chart of ideas, using students own words and examples. Accept all students' responses.

Then distribute S-1 to students and have them complete the page independently.

Wrapping Up

Plan to refer to the chart periodically as you work through this unit, encouraging students to examine their ideas in light of new learning.

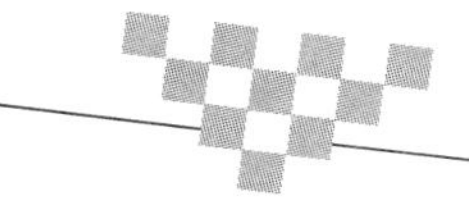

TEACHING TIP

INFORMAL ASSESSMENT

It is helpful to have students write their ideas in their journals before holding a discussion as a group. Some students may not share an idea unless they hear someone share it first. By having them write in their math journals first, students are more likely to share a wide range of insights.

What Really Happened

Initially, students had a hard time talking about what addition is all about. However, as the discussion progressed and they realized that this was a question with many answers, ideas began to flow. Students pointed out that when they add, they get a bigger number and when they subtract, they get a smaller number. In their examples, students most often mentioned using adding in a store or in math class or when doing homework.

Examples of subtraction most often related to taking away things. One student suggested to subtract to share things. Students did not naturally use subtraction in situations to compare quantities.

Gregory, a prolific writer who invents spelling as needed, is a first grader. He explained his addition strategies after a lengthy essay explaining that addition is hard but he tries his best. He writes, "I count with my fingers and sometimes I know the answer. I know that 6 + 6 = 12 and I know that 3 – 3 = 0, and I know that 10 x 10 is = 100. When we have math, that is when we use adding."

Sharing Ideas and Strategies

Students in this second-grade class had little experience with writing about math. Their responses were very short and often incomplete. We wrote follow-up questions in their journals. Then, after they responded, we shared ideas as a class.

Students' ideas about addition and subtraction were very similar. Addition means putting things together. Subtraction means taking numbers apart or taking away.

Substraction is taking afew from a lot.

In follow-up discussion, Brian agreed that you could also take a lot away from a lot. Then you wouldn't have much left over.

Alicia writes, "Subtraction is like a mystery like 9 – 9 = 0." Alicia remembers problems with missing numbers where she had to find what number you take away from 9 to get zero.

sudstrlction is Like misesti Like 9 -9 0

Chelsea
What is substraction?
It is something that takes away anumber from anoter number or shape.

Dear Chelsea,
I like your explanation.
I wonder exactly what you mean by shape. Please explain this idea further!

Well, Its like if there is 3 circles and you take away there would be 2 left.

Chelsea mentioned shapes as a part of subtraction. When we asked her for more detail, Chelsea explained a model for subtraction from her textbook.

Lesson 2 Skip Counting by Tens

Materials

- tagboard sentence strips
- marker

Advance Preparation

Prepare ten tagboard sentence strips, each 36 inches long. Mark each strip into 3" segments. Do NOT fill in the numbers ahead of time.

This lesson is an ongoing activity over several days. Add number strips to your collection until your students have skip counted from every number between 0 and 10. Then vary the activity by using starting numbers like 23 or 47 or 71. When you have number strips for every starting number from 0–9, practice counting backwards by tens from different numbers. Call on a different volunteer each day to choose a starting number and to decide whether to skip up or skip back by tens.

NOTE: Your students can help fill in the counting strips over several days.

1	2	3	4	5	6	7	8	9	10
11	12	13	14	15	16	17	18	19	20
21	22	23	24	25	26	27	28	29	30
31	32	33	34	35	36	37	38	39	40
41	42	43	44	45	46	47	48	49	50
51	52	53	54	55	56	57	58	59	60
61	62	63	64	65	66	67	68	69	70
71	72	73	74	75	76	77	78	79	80
81	82	83	84	85	86	87	88	89	90
91	92	93	94	95	96	97	98	99	100
101	102	103	104	105	106	107	108	109	110
111	112	113	114	115	116	117	118	119	120
121	122	123	124	125	126	127	128	129	130

Objectives

- to discover number patterns when skip counting from numbers other than zero
- to build a number chart from skip-counting strips

Getting Started

Explain to students that today they will skip count. Ask them what they already know about skip counting. Skip count in unison with your students by twos, fives and tens to 100 or beyond.

Then use a sentence strip to record vertically the numbers you say when you skip count by tens.

Skip Counting by Tens

Explain that today students are going to skip count by tens from 5. Begin by counting chorally. Listen to the strength of student voices to assess for comfort levels. Repeat the sequence several times so that students build confidence with the pattern.

Then record the numbers vertically on a sentence strip. Post the sentence strip. Continue in the same way over several days, counting from other numbers, until you can make a complete hundred chart.

2	5	10
12	15	20
22	25	30
32	35	40
42	45	50
52	55	60
62	65	70
72	75	80
82	85	90
92	95	100
102	105	110
112	115	120
122	125	

Wrapping Up

Have students examine the posted sentence strips for counting by tens from zero and from five. Ask questions like, "What do you notice about skip counting by ten starting with five? What patterns do you see?" Elicit that there are number patterns. (There is always a five in the number. It is always in the same place. The digit in the tens place goes up by one each time.)

Extension

Extend the skip counting past 100. Add strips to show the pattern. Then have students predict when the digit in the hundreds place might change.

Lesson 3 USE HUNDRED CHART PUZZLES

Materials

- hundred chart from previous lesson
- **Number Puzzle** and **Number Puzzle Pieces,** S-2 and S-3
- envelope
- overhead transparency
- overhead projector

For **Extensions/ Homework**

- paste, glue, or tape
- **Hundred Chart Puzzlers** and **More Hundred Chart Puzzlers**, S-4 and S-5

Advance Preparation

To create a hundred chart with your students, see **Skip Counting by Tens**, page 25.

Make a transparency of **Number Puzzle**, S-2, and of **Number Puzzle Pieces**, S-3. Cut puzzle pieces apart and store in an envelope.

Make enough copies of **Number Puzzle,** S-2, and **Number Puzzle Pieces,** S-3, for pairs or small groups of students to use.

Make a copy of **Hundred Chart Puzzlers**, S-4, and **More Hundred Chart Puzzlers**, S-5, for each student.

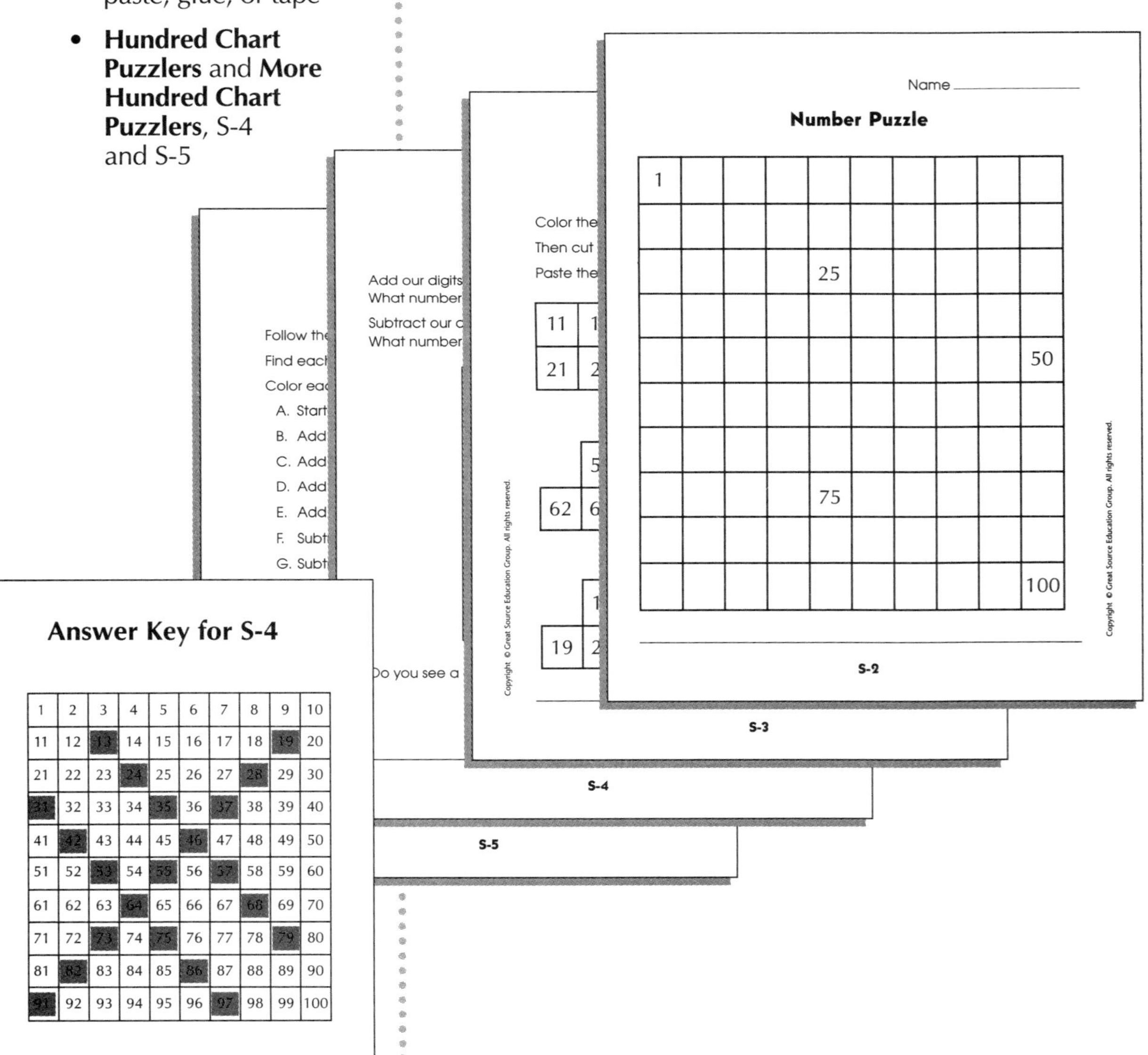

1	2	3	4	5	6	7	8	9	10
11	12	13	14	15	16	17	18	19	20
21	22	23	24	25	26	27	28	29	30
31	32	33	34	35	36	37	38	39	40
41	42	43	44	45	46	47	48	49	50
51	52	53	54	55	56	57	58	59	60
61	62	63	64	65	66	67	68	69	70
71	72	73	74	75	76	77	78	79	80
81	82	83	84	85	86	87	88	89	90
91	92	93	94	95	96	97	98	99	100

Objectives

- to order clusters of numbers in a hundred chart
- to develop spatial visualization using a hundred chart

Getting Started

Call students' attention to the hundred chart created in lesson 2. Ask questions like, "What does this chart tell us? How can we use this chart?"

If students don't spontaneously talk about patterns in the chart, you may want to ask additional questions such as:

"What do you notice about the numbers?"

"What patterns do you see?"

Elicit that there are horizontal and vertical number patterns.

Display the hundred chart on S-2 and one of the puzzle pieces from S-3. Ask students what they notice about the puzzle piece. Then ask them where they think this puzzle piece fits on the chart. Explain that today students will be working with a partner to put together number puzzles of their own.

Use Hundred Chart Puzzles

Have students work with a partner or in small groups. Distribute **Number Puzzle**, S-2, and **Number Puzzle Pieces**, S-3, to each group. Have students color and cut out the pieces on S-3. Explain that students will work together to decide where each puzzle piece should go and explain why. When they agree on a placement, they may paste the piece in place and get another.

Wrapping Up

Have volunteers post their puzzle solutions. Discuss each as a class. Decide whether students agree with the placement of the puzzle pieces. Have students explain their reasoning.

Extensions/Homework

Assign pages S-4 and S-5 as in-class activity extensions or for homework. If you assign for homework, make time during the next math class to display and discuss results.

Answer Key for S-5

1	2	3	4	5	6	7	8	9	10
11	12	13	14	15	16	17	18	19	20
21	22	23	24	25	26	27	28	29	30
31	32	33	34	35	36	37	38	39	40
41	42	43	44	45	46	47	48	49	50
51	52	53	54	55	56	57	58	59	60
61	62	63	64	65	66	67	68	69	70
71	72	73	74	75	76	77	78	79	80
81	82	83	84	85	86	87	88	89	90
91	92	93	94	95	96	97	98	99	100

TEACHING TIP

INFORMAL ASSESSMENT

You may want to use this activity as an assessment. Work with small groups and interview students about their thinking as they consider where to place their puzzle pieces.

Observe for familiarity with the structure of a hundred chart, for comfort reading numbers out of sequence, and for abilities to sequence and order numbers that skip.

Lesson 4 Keep the Difference

Materials

- **Keep the Difference**, S-6
- ten pairs of two number cubes
- ten pairs of custom number cubes (See **Advance Preparation.**)

Advance Preparation

Assemble these supplies:

- blank wooden or plastic cubes
- paper stick-on labels
- fine-tipped permanent marker

To create the custom number cubes, stick a label on each face of the cubes. For each pair of cubes, mark the numerals 1–6 on one, and 7–12 on the other.

Make a copy of **Keep the Difference**, S-6, for each student.

Name ________________

Keep the Difference

I played with ______________________________

I rolled	My partner rolled	Difference
3 + 6 = (9)	6 + 6 = (12)	12 – 9 = 3

S-6

Objectives

- to review basic addition and subtraction facts
- to find patterns and organize subtraction facts
- to explore combinations and probabilities related to a game

Getting Started

Assign partners or have students choose a partner. Then explain that you are going to play an addition and subtraction game called "Keep the Difference." Ask a volunteer to play several demonstration rounds with you, first using the custom pair of dice and then using the two standard dice.

Demonstrate how to play the game, following the rules at right. Record each play.

Keep the Difference

When students understand how to play, distribute S-6 and one set of number cubes to each pair of students. Be sure there are at least two of each kind of number cube in use. Then circulate as students play the game and observe for facility with basic facts.

Wrapping Up

Gather the class for discussion. Have volunteers describe what they noticed as they played. Be prepared for some great discoveries because the custom number cubes are "loaded." This discovery will lead to tomorrow's lesson.

How to Play "Keep the Difference"

You need:

a partner
a set of number cubes
a recording sheet, S-6

1. Player 1 rolls the number cubes and records the sum.
2. Player 2 rolls and records the sum.
3. Players figure out the difference between the sums.
4. The player with the lower score gets to "keep the difference."
5. The game is over when one player has earned a total of 25 by keeping the difference.

What Really Happened

This game was very popular. As students played, we observed students' addition and subtraction strategies, taking notes in our teacher's journal. We noticed a variety of addition strategies.

A few students "just knew" the sums, having memorized the combinations earlier.

Some students counted on from one number to find a sum. Of this group, a few needed to use touch counting in order to maintain one-to-one correspondence. Others could count in their heads or on their fingers.

A large number of students used a pull-apart method to create easier combinations. For example, Brandon saw that 11 + 6 was like 10 + 7. Charles saw that 8 + 5 was the same as 8 + 2 = 10 plus 3 more makes 13. Samantha used a similar strategy for 8 + 6, thinking, "8 + 3 = 11 and 3 more is 14." To add 12 + 6 Michael took 2 from the 12 and added that to 6 to make 8. Then he knew that 8 + 10 is 18.

To find the differences between the two sums, students most often used addition. They figured out how many more they would have to add on to equalize the numbers.

After about twenty minutes of play, several groups of students seemed perplexed. We stopped the activity and gathered the class for discussion.

"Something is wrong here. We keep coming up with 0, 1, or 2. We can't get any other differences."

We asked the class to share their recording cards. "What sums do you seem to get most often?"

We decided to make a chart of the possible sums and discovered that the groups using the custom number cubes most often got 12, 13, or 14. The group using standard cubes seemed to get a greater variety of sums.

The group with custom number cubes decided to change the rules of the game. "Instead of adding first, let's just each roll cubes and subtract and see what kinds of answers we get." This suggestion resulted in the graphing activity described on pages 32–34.

Sharing Ideas and Strategies

Students wrote letters to the "Games Department," sharing their ideas for improving the game. Shaylan and Erin wrote about two ways to improve the game.

I liked the game but I think I got a way that I can make the game better we could not yous the cubs and ferger it out in our brains.

Erin suggested playing the game without using number cubes. "We could figure it out in our brains."

It Was Lots of fun. We cude add two more Dice. If We yous 2 more Then we cude come up with a Hire Numre.

Shaylan suggested using more than two number cubes at a time so that your rolls would add up to higher numbers .

Organize and Graph Data

Materials

- large sheet of paper
- stapler or tape
- number cubes 1-6 and 7-12
- tagboard

Advance Preparation

Create game cards by cutting an 8 1/2" by 11" sheet of tagboard into eighths. Make ten cards for each student.

Objectives

- to practice subtraction
- to make a graph of basic subtraction facts
- to explore ideas of probability

Getting Started

Have students review their experiences playing "Keep the Difference." Then tell them that today they will explore probability by recording all the differences between rolls of the number cubes 1–6 and 7–12. Explain that they will work in pairs and each partner rolls one number cube; then players compare their numbers. They record each subtraction sentence on a card. The first pair to reach 36 subtraction sentences raises their hands.

Organize and Graph Data

Explain that you would like to collect student data to look for patterns in addition and subtraction. Have each pair find a way to organize and graph their cards. Suggest that students focus on the differences recorded on the cards when deciding how to organize them.

When partners have agreed on a way to organize their cards, have them tape or staple them to a large piece of paper.

When students have finished, ask each pair to describe their sorting system. Have students compare their results. Ask students to hypothesize about the different outcomes.

Wrapping Up

Discuss the shape of each graph. Elicit that some differences are more likely than others, regardless of which set of number cubes you use.

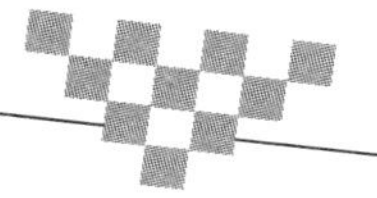

TEACHING TIP

USING WRITING PROMPTS

If students are having trouble describing the experience, you may want to post one or more of these writing prompts:

Which number cube did each of you use?

What did you notice about your subtraction cards?

How did your team organize the subtraction cards?

What Really Happened

Students made organized lists and graphs of all the differences between their rolls of the number cubes 1–6 and 7–12. Students found that rolling one of each kind of cube resulted in the most interesting possibilities.

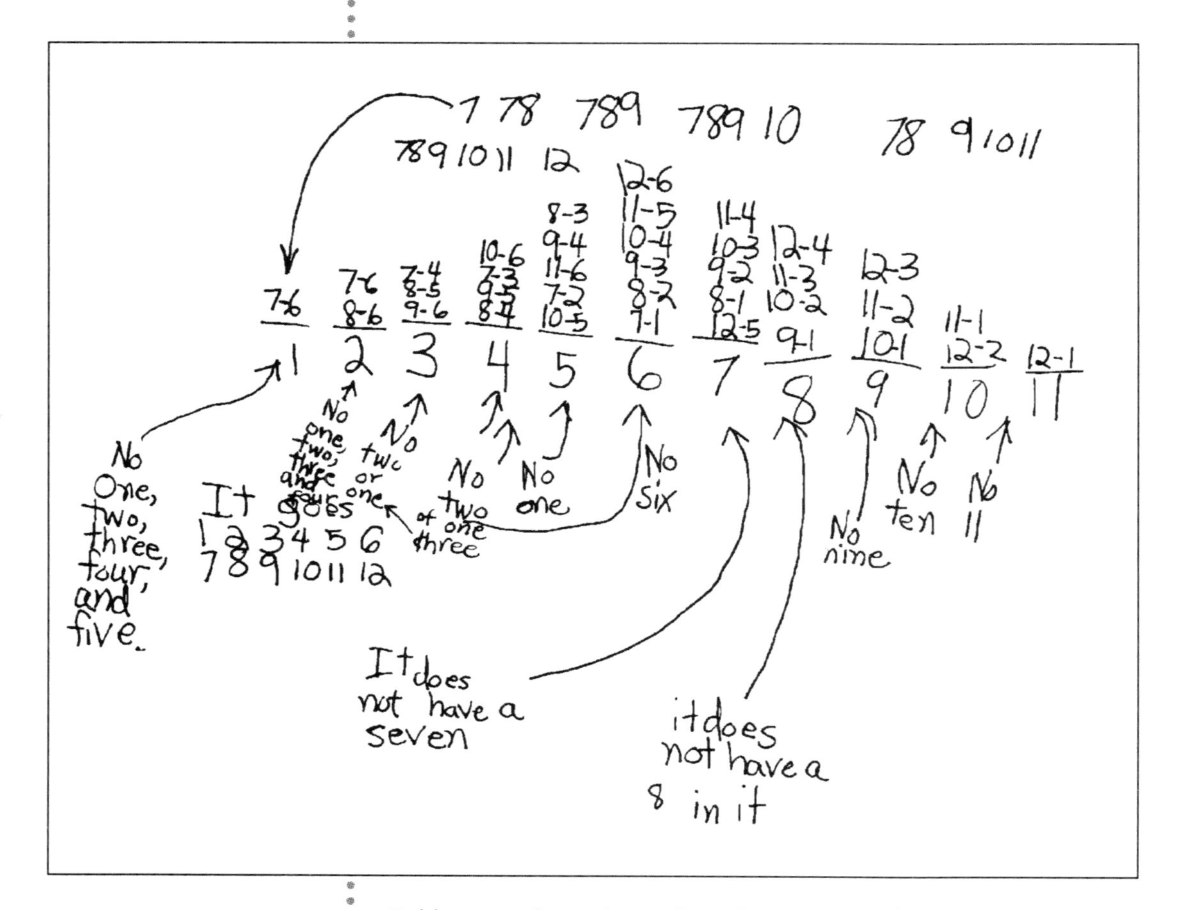

Bobby started out placing his subtraction problems in random order. By the time he reached differences of 5, however, he saw a pattern and began to organize the subtraction problems by the minuends ("You always start with the biggest number first," he explained).

Other students in the class saw a pattern in Bobby's chart. "The first number goes up by one and so does the second number," Angie commented. For each possible difference, Bobby noticed that some subtraction combinations were not possible with these number cubes.

MOVING BEYOND BASIC FACTS

About Teaching Place Value

First- and second-grade students have powerful intuitions about how two-digit and three-digit numbers work. For example, if you ask how many gumballs there are if you have 34 and I have 20, students will use ideas of place value in their reasoning. Their typical response is: "50 because you just add 3 and 2 which is 5, but the numbers are bigger so the answer is 50, not just 5. Then you have 4 more and that makes 54."

This highly successful computation strategy shows that students have a powerful sense of place value and use it spontaneously to make connections between several experiences: observing patterns as they rote count to 100, using patterns as they skip count by tens, composing and decomposing numbers in some way. The ability to make these connections is strongest in the oral/aural mode as seen in the following example. I asked a group of first-graders: "If Juliana has 202 marshmallows and I have 309, how many do we have together?" A majority of students gave the following explanation: "2 + 3 is 5 and you are talking about hundreds so that is 500. Then you have two more to make 502 and nine more which is 11 so the answer is 511." No one asked to see the numbers written down.

Why do students who have such powerful computation skills "fail" with multi-digit addition and subtraction? I believe that traditional ways of teaching computation short-circuit their number sense by imposing standard mathematical notation and set procedures that do not relate to the ways students think about numbers.

In my experience, directly teaching place value concepts as a way of understanding addition of larger numbers was confusing for many students and counter-productive for most. Instead, visual models such as acting out stories and analyzing ten-frames encouraged students to visualize larger numbers, and to see patterns for combining and decomposing numbers to add and subtract.

Ten-frame Math 1

How many hearts do you see? How did you figure it out?
How many umbrellas do you see? How did you figure it out?
How many squares are filled? How many squares are empty?

112

When presented with this visual model (page 112), none of my students had trouble adding 25 and 17. Their strategies revealed sound place value sense and sound number sense. Most importantly, their mental math was insightful and their sums were accurate.

When asked to record addition mathematically, many students seem to lose their number sense. In the stars and suns picture, Brian said that he saw 40 pictures in all. He explained, "7 goes with the 23 to make 30, and 10 more makes 40." However, when asked to write an equation for the picture, he wrote, "17+23=31." I asked him to explain how his equation worked. He explained, "7 and 3 is 10 so you put down the 0 and carry the 1. 10 and 20 is 30 and one more is 31."

At this point, I asked Brian whether there could be more than one right answer. "No, because you have what you have," he replied. Then I asked him to work on his own and figure out which answer was right, 40 or 31.

Ten-frame Math 5

How many stars do you see? How many suns do you see?
How many squares are filled? How many are empty?
Can you make up a number sentence about the empty boxes?

116

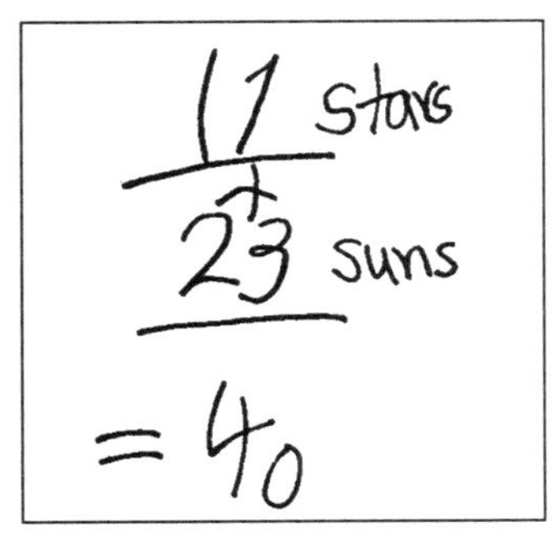

When Brian was ready to prove his answer to me, I noticed that he used the visual model to explain that you could add 10 and 20 first, then 7 and 3 and then add that all up, so 40 had to be the right answer. Here he used an equation only for recording but did all the figuring in his head. Clearly, rushing Brian to use mathematical notation and set procedures got in the way of his understanding of place value in addition.

Ten-frame activities are introduced on pages 38–41. After introducing ten-frame math, we suggest that you include a ten-frame activity as a five-minute warm-up on a daily basis. Copymasters are provided on pages 112–117. These visual models support students' intuitions about combining numbers and build success with multi-digit addition and subtraction.

Most importantly, ten-frame activities give you an opportunity to listen as students explain their reasoning. By listening to students' ways of thinking, we can help them deepen their number sense, extend their operations sense to larger numbers, and encourage them to build sound logical foundations for dealing with computation.

Lesson 6

Introduce Ten-frame Math

Materials

- ten counters
- **Ten-frame Workmat**, page 110
- **Ten-frame Math 1**, page 112
- overhead transparencies
- overhead projector

Advance Preparation

Prepare transparencies of a **Ten-frame Workmat**, page 110, and **Ten-frame Math**, page 112. Plan to present one or two ten-frame math activities as a warm-up for a week or more to build students' confidence with addition and comparative subtraction using pictorial models.

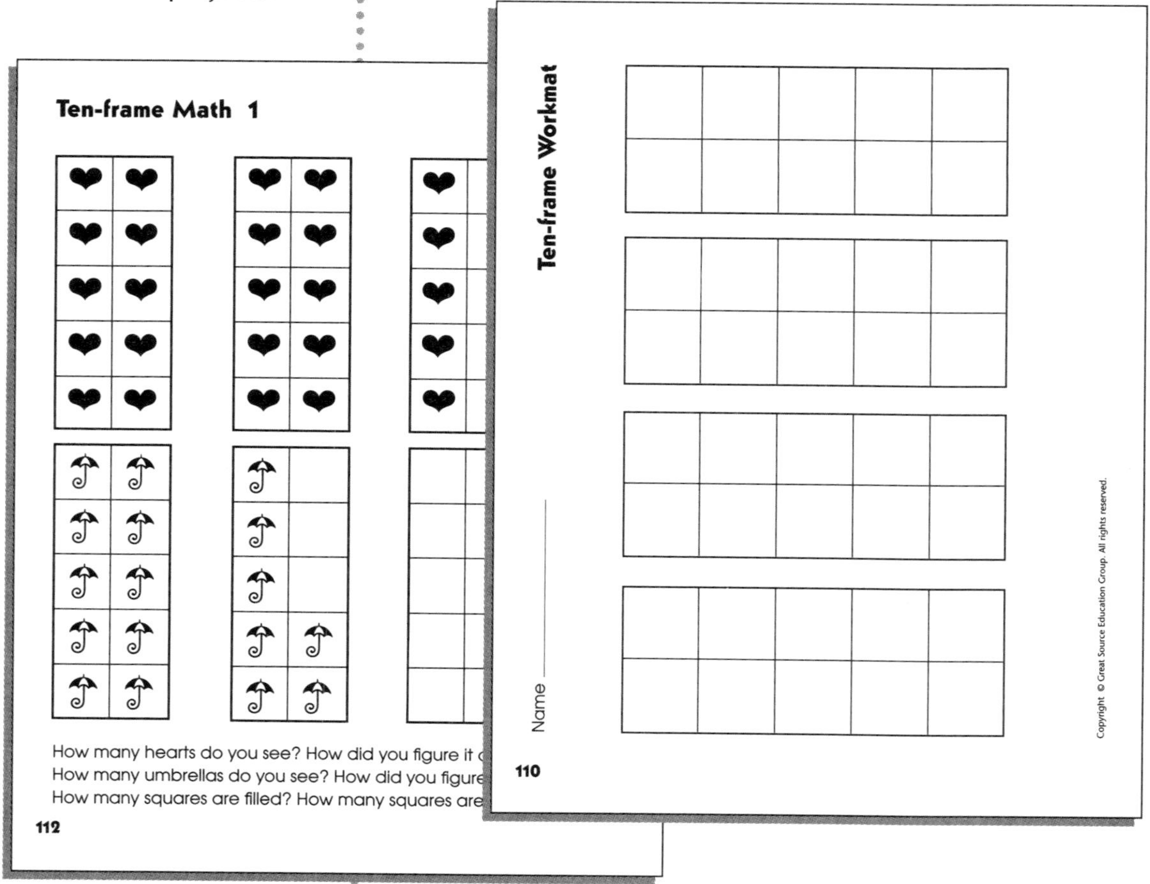

Objectives

- to visualize addition and subtraction
- to develop mental math skills
- to explore adding two-digit numbers mentally
- to compose and decompose tens within two-digit addition

Getting Started

Display a ten-frame workmat on the overhead. Ask students to describe what they see. Elicit that the ten-frame workmat shows two rows of five boxes each.

Introduce Ten-frame Math

Introduce ten-frame math as described on page 40.

After introducing ten-frame math and discussing students' reasoning, plan a ten-frame activity as a daily warm-up for the next week or two. You will notice that your students develop facility with composing and decomposing tens as they exercise their intuitive mental math skills. Be prepared for a variety of computation strategies.

Once again, display the ten-frame workmat on the overhead.

Place 5 counters in one vertical row. Ask, "How many counters are there? How many more counters do I need to make a sum of 10?"

Continue with other facts to 10. Point out that one row is filled up before filling up the other row.

As students share different ideas for how to use the ten-frame pictures, you will notice a marked improvement in all students' facility with mental math.

Display **Ten-frame Math 1**, page 112. Ask students to estimate the total number of hearts in the frames. Then remove the display and ask them to guess how many hearts they saw. Elicit a guess from each student.

Ten-frame Math 1

How many hearts do you see? How did you figure it out?
How many umbrellas do you see? How did you figure it out?
How many squares are filled? How many squares are empty?

112

Display the overhead again so that only the hearts are showing. Ask, "How many hearts do you see?" Encourage students to share their counting strategies.

Next, display the whole page and ask students to guess how many umbrellas are in the frames. Continue in the same way as before, eliciting an estimate and then giving students time to check their estimates.

Finally ask, "How many umbrellas and hearts do you see? Think about how you would find the answer."

After students have shared their thinking, demonstrate how to write an addition equation for hearts and umbrellas. However, do not ask students to write equations at this time.

What Really Happened

Students had no trouble describing their observations when we showed them a ten-frame workmat (page 110) and asked what they noticed about it.

We then displayed **Ten-frame Math 1** (page 112) to the whole class. "Look at this sheet and see how many hearts you can see," we said.

After a few moments, we covered the paper and asked students how many hearts they thought there were. Panic swept the room.

"I need to see it again. I haven't finished counting!"

After reminding students that this was a guessing game and they would get a chance to see the page again, they were more willing to take a guess. Every student was invited to give their guess.

Responses varied. Brian thought there might be 16. Natasha suggested that there might be 45. Robert and David said they had seen 24. Matt thought there might be 15. Emma suggested 14. We listed all estimates.

"Now I'm going to show you the pictures again. Check to see how close your estimate was."

We held up the paper, covering all but the hearts. When students were ready, we called on each child in turn to tell how many hearts they saw and how they figured it out. Strategies varied.

Robert had carefully counted every heart. However, his explanation showed that he was thinking about this problem in several different ways. "I added 10 plus 10 makes 20, plus 5 more makes 25."

Kate said, "First I skip counted 10, 20, and then just counted 21, 22, 23, 24, 25." Brennan also used skip counting plus counting on.

Natasha said, "I just counted them all up by ones."

Everyone agreed that there were 25 hearts except Emma who said she saw 23. We accepted these responses. Students were eager to try some more."Let's estimate how many umbrellas there are."

We followed the same procedure, first showing the page briefly to encourage estimation. As students examined the paper, we observed their counting strategies, noticing that some students were counting each individual image. Others seemed to be counting on from ten.

This time, almost all guesses were exactly correct. I asked students to explain how they had estimated.

Juliana said, "I know that one part of the ten-frame is 5 because 5 + 5 is 10 and the two parts are the same. So I counted 10, 15, and then put in the two more, 16, 17 umbrellas."

Matthew saw the same relationship but explained it differently. "I just counted by fives because each column is five so it's 5, 10, 15, and 2 more makes 16, 17."

Finally, we asked students how many hearts and umbrellas they saw. Students shared addition strategies and found that there were 42 pictures in all.

Then I asked one more set of questions."Are there more hearts or more umbrellas?" Students agreed that there were more hearts.

"How many more hearts are there?" Students seemed stumped so I rephrased the question. "If every heart gets an umbrella, how many hearts will not have an umbrella." After thinking about this situation, students counted "leftovers" to figure out that there were eight more hearts than umbrellas.

We explained that we would do more ten-frame math over the next few weeks. Students were enthusiastic about trying more of these activities.

Lesson 7

Group and Count Objects

Materials

- ***Swimmy*** by Leo Lionni
- **Ten-frame Math 2**, page 113
- cups
- blue construction paper (or paper plates and blue crayons)
- goldfish crackers (about 15 per student)
- overhead transparency
- overhead projector

Advance Preparation

Prepare an overhead transparency of **Ten-frame Math 2**, page 113.

Prepare a cup for each student by randomly placing 10-15 goldfish crackers in each one.

Cut out paper plate shapes from blue construction paper. Or, you may wish to have students color paper plates using blue crayons.

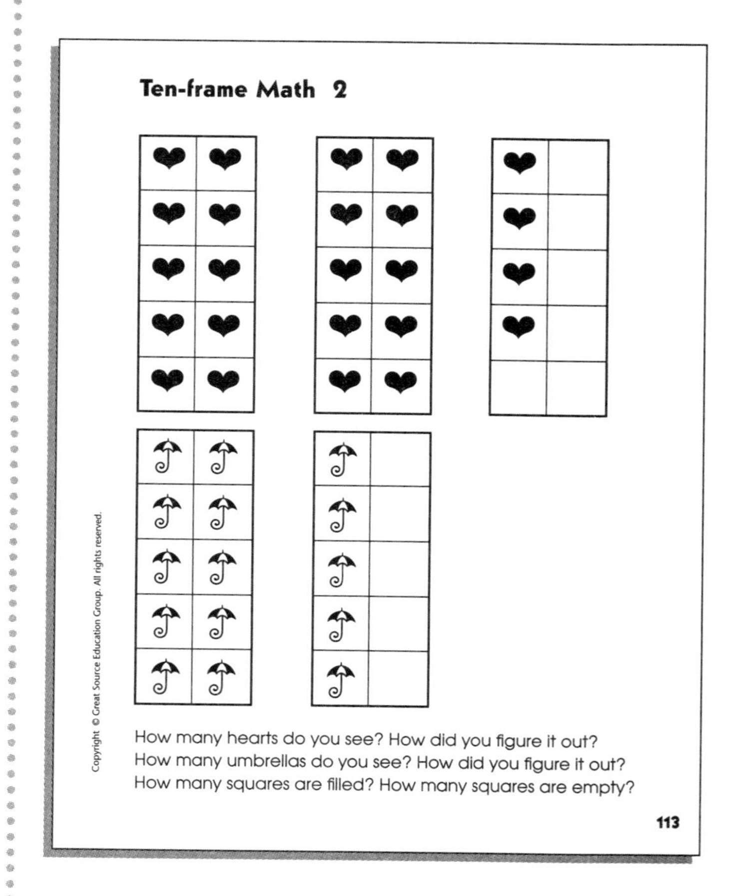

Objectives

- to make visual estimates
- to arrange objects to make them easy to count
- to assess informally for grouping and counting strategies

Getting Started

As a warm-up, present a **Ten-frame Math** activity as described on page 40, using a transparency of **Ten-frame Math 2**, page 113. Then teach the day's lesson.

Explain to students that today they will read a story about a fish named Swimmy. Display the cover of the book and ask students to describe what they see. Then read the book with the students.

After reading, discuss the story. Turn to the last page and ask students about how many friends Swimmy used to form this big fish. Elicit that the actual number is hard to count.

Group and Count Objects

Distribute cups of fish crackers and paper plates to each student. Have students estimate how many crackers are in their cups. Then have them spill out the crackers and arrange them in some way for counting.

Circulate and observe as students work. This is a good time to conduct individual interviews about counting strategies. Encourage early finishers to pair up with another student and figure out how many fish crackers they have together. Observe how students apply problem solving and addition strategies.

Wrapping Up

Have students share their counting strategies. Then have them compare their estimates to the actual number of fish that were in their cups. Finish the activity by saying, "Everyone may now eat one fish cracker. How many do you have left? How do you know?"

"Now everyone may eat five more fish crackers. How many do you have left now? How do you know?"

"If you ate half of the crackers you have left, how many would that be? How do you know?"

What Really Happened

Many students in this first grade class recognized the story book when we presented it. They shared their wisdom and then were ready to listen. "I read that book in kindergarten!" "I know that story!" "My dad's name is Leo!" As we started reading, we suggested that students pay attention to how the illustrator created the pictures.

As we read, we asked number questions. "How many fish do you think the tuna ate?" Students responses varied widely, "100, lots, 20, 200."

After we finished reading, we asked students what they noticed about the illustrations. "He uses sponge painting," Juliana shared.

"There are lots of fish to make the big fish. It's bigger than the tuna," Zack noticed.

"Look how small Swimmy is compared to the big fish," Robert added.

Then we asked students to look again at the picture of Swimmy and his friends. "How many fish do you suppose there are to make such a big one?" We asked each student for an idea. Estimates ranged from 50 to 8 million to "one zillion 42."

Then Natasha declared that she wasn't about to try to count those fish.

"Why not?" we asked.

"Because it is too hard. There are too many, you can't see them, you'd have a hard time figuring it out."

This challenge inspired some problem solving.

"You could count by tens," Emma offered.

"You could divide the picture down the middle," Brian said. "Then you count the part in one half and then the part in the other half and just put them together." Several students agreed that Brian's was a great idea.

Natasha offered another variation. "You could cut the big fish into pieces and count the fish in the pieces and then put the numbers together."

"You could ask the author how many he made," Juliana said practically.

We wrapped up the discussion with more questions.

"Could we count the fish by ones?" "Yes!"

"Would we want to count the fish by ones?" "No! You'd get too tired and you'd get mixed up."

We then asked students to sit in a circle and get ready for a fish activity. Two student helpers distributed blue paper plates and paper cups containing random numbers of fish crackers (between 10 and 15 each).

"Look into your cup. Without touching your crackers, estimate about how many you think there are. Tell your neighbor your estimate." We gave students a few moments to count. When we heard murmured numbers, we asked them to tip out the crackers onto their plates and count them.

"Now arrange your crackers to make your fish easy to count." No one made easy arrays! As students made creative fish shapes, we circulated and asked students about their counting strategies.

Sharing Ideas and Strategies

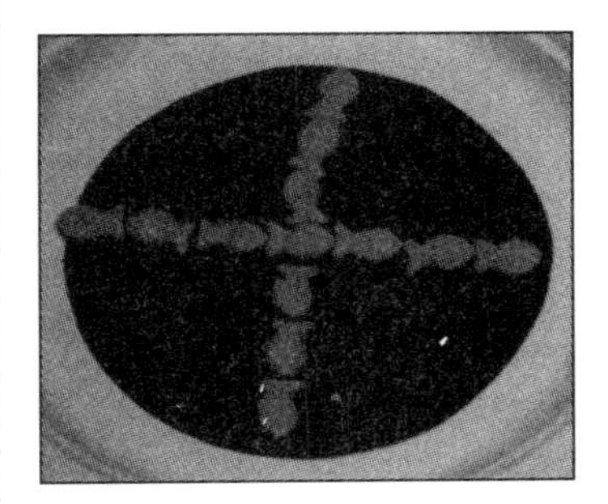

Brian had made a sea star out of his 13 crackers. His sea star was nicely symmetrical with four tentacles and the odd cracker in the middle to anchor his pattern. When asked to describe his counting strategy, he counted by twos. "It makes 13. That's an odd number," he concluded.

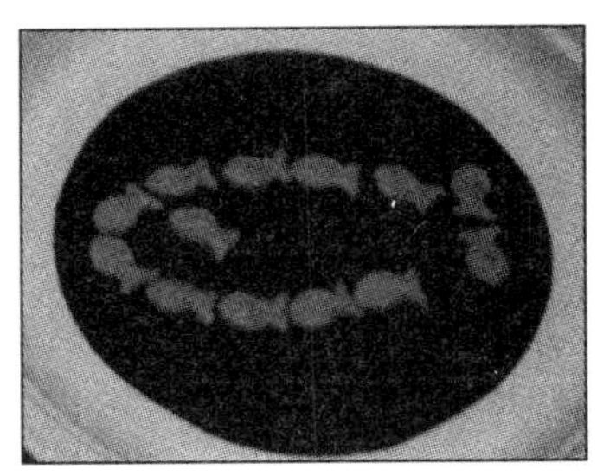

Several children had made fish outlines. Emma found it easiest to separate out ten and then add on the rest even though it made a hole in her fish outline. She included Swimmy as the eye of her fish. "I have ten crackers and three more makes thirteen."

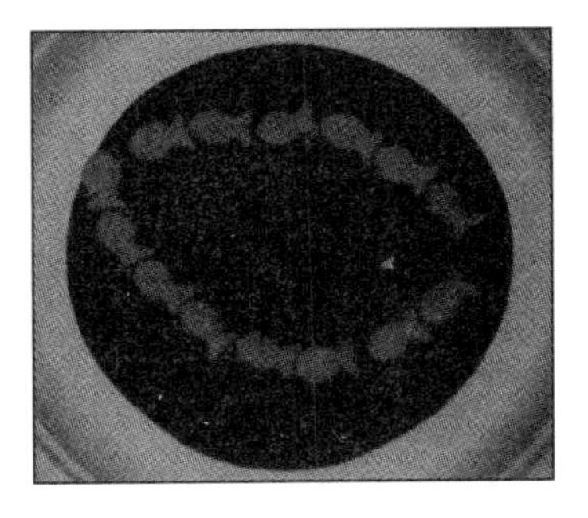

Natasha made a fish outline. "I divided my crackers in half. There are 8 fish in one half and 6 in the other. That's 14 crackers. I just know. I can count by ones to prove it." Natasha counted up from 8 to show her method.

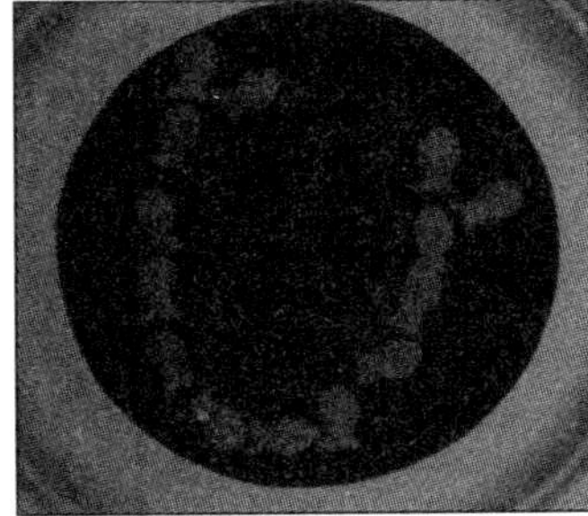

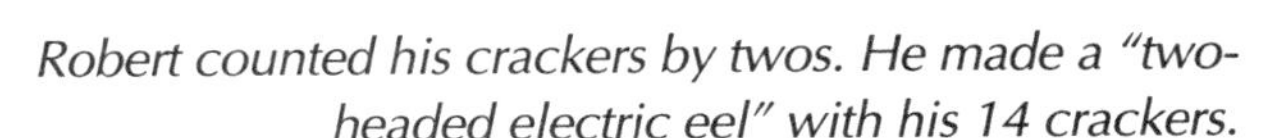

Robert counted his crackers by twos. He made a "two-headed electric eel" with his 14 crackers.

Lesson 8 Estimate Fish Area

Materials

- **Swimmy** by Leo Lionni
- **Estimating Fish Area**, S-7
- **Ten-frame Math 3**, page 114
- goldfish crackers (ten for each student)
- small paper cups
- overhead transparency
- overhead projector

Advance Preparation

Prepare a transparency of **Ten-frame Math 3**, page 114. For each student, prepare a small paper cup with ten goldfish crackers in it .

Make copies of **Estimating Fish Area**, S-7 for the students. Also, create a transparency of S-7 to use to demonstrate this activity.

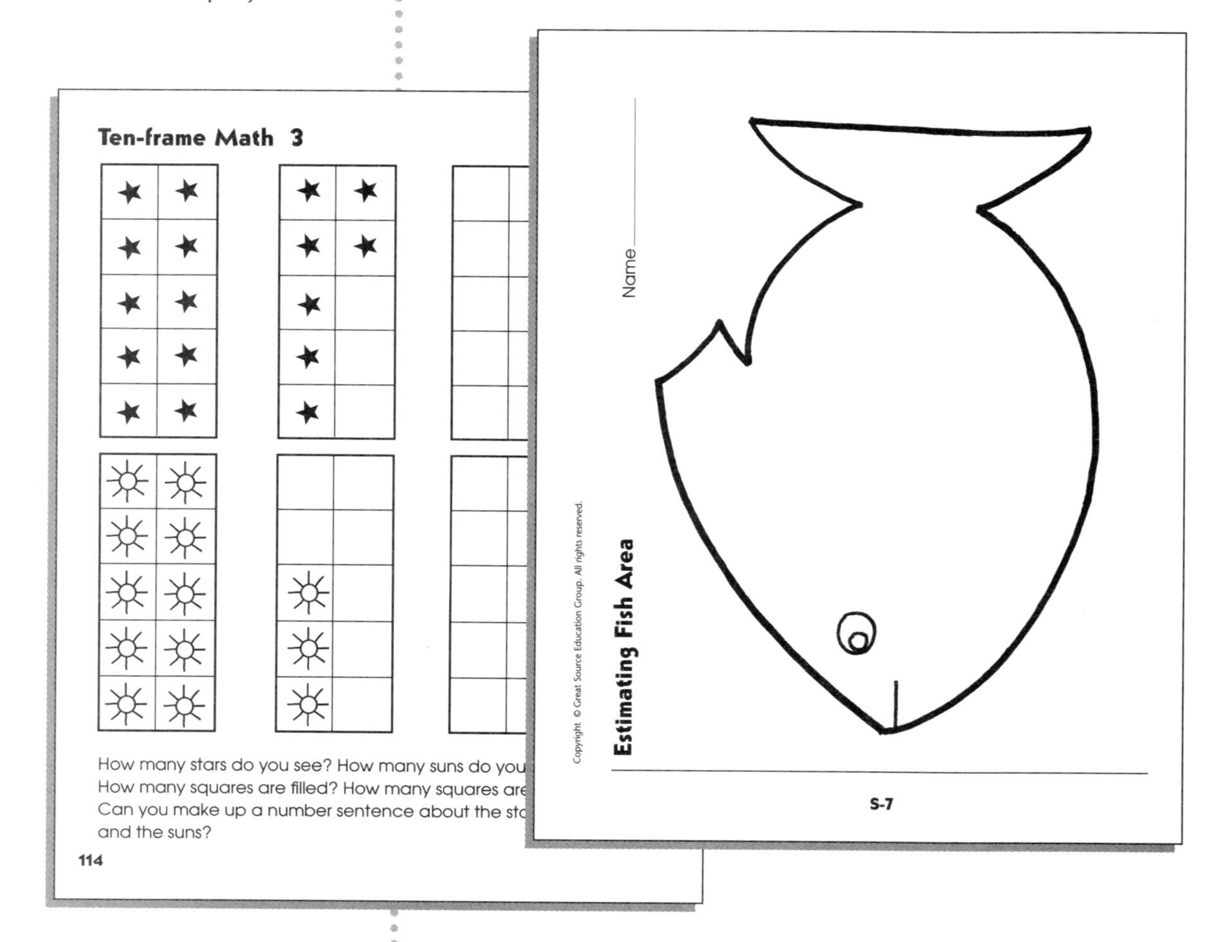

Objectives

- to estimate quantities using benchmarks
- to develop number sense for counting by tens

Getting Started

As a warm-up, present a **Ten-frame Math** activity as described on page 40, using a transparency of **Ten-frame Math 3**, page 114. Then teach the day's lesson.

Reread ***Swimmy.*** At the end of the story, ask, "How many fish are there in this big fish?" Then explain that today students will be estimating how many little fish it might take to make a big fish.

Estimate Fish Area

Display a transparency of **Estimating Fish Area,** (S-7) and one goldfish cracker. Ask, "How many of these do you think it will take to cover the big fish?" Record students' ideas. Then use 10 goldfish to cover an area of the fish outline. Invite students to revise their estimates. Accept revisions and record changes. Ask students how you could use the same 10 fish to cover more of the Big Fish. Elicit that you could trace around the outline of the area you have covered and then move your 10 goldfish to a new area and count some more.

Continue in the same way until the fish is covered. Then discuss results. Ask questions like, "How many fish crackers did we use? Which estimate was closest? Could we have used more or fewer crackers?"

Distribute a copy of S-7 and 10 goldfish crackers to each student. Instruct students to estimate and record how many little fish they think it will take to cover the big one. Then have students use the strategy of cover and record to verify their estimates. Allow them to revise their estimates each time they use the 10 goldfish to cover more of the fish.

Circulate and observe. Invite early finishers to color the areas they have traced in different colors to make a beautiful fish.

Wrapping Up

Invite volunteers to share and compare their findings.

What Really Happened

We found that some students were very reluctant to offer an estimate. They wanted their estimate to be exactly right! Sometimes a leading question was helpful. We asked, "Do you think that you will need more than 50 fish or fewer than 50 fish?" With that prompt, students decided it would be fewer than 50 and were then willing to make a first guess.

After we each had laid out ten fish crackers on our big fish, we outlined the area the crackers covered. We then moved the fish crackers aside and recorded a 10 in the area. We asked the group whether they would like to change their estimates based on the area 10 fish had covered. We noticed that some children analyzed the area left to cover and changed their estimates.

The following are some examples of individual students' estimates and strategies:

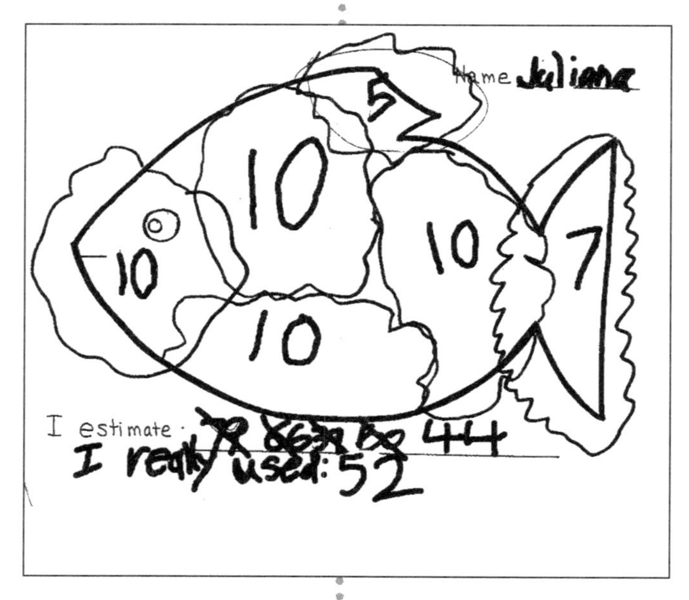

We asked Juliana what she thought of her estimate after she had covered two ten-fish areas. "What do you mean?" she asked.

"Look at the areas you have outlined. How many fish did you use?"

"Two tens, so 10, 20."

"Is that about half your fish yet?"

"Well, not really. Half would be further over here."

"So how many more fish do you think you might need?"

"I don't know. I guess I'd need at least 20 more."

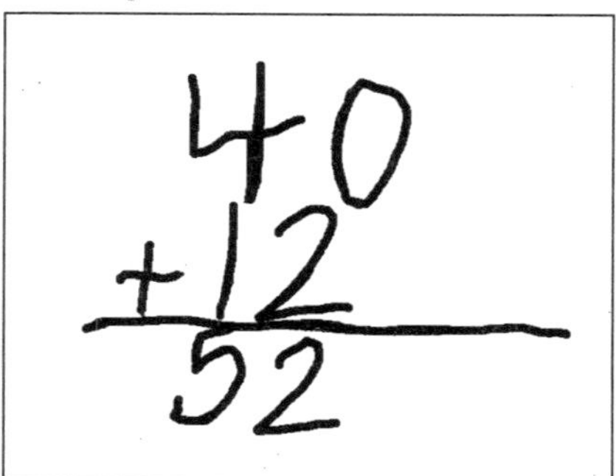

Juliana created four areas of 10 fish each and two areas with fewer than 10 fish each. She wrote an equation with two addends and solved it using the standard addition algorithm.

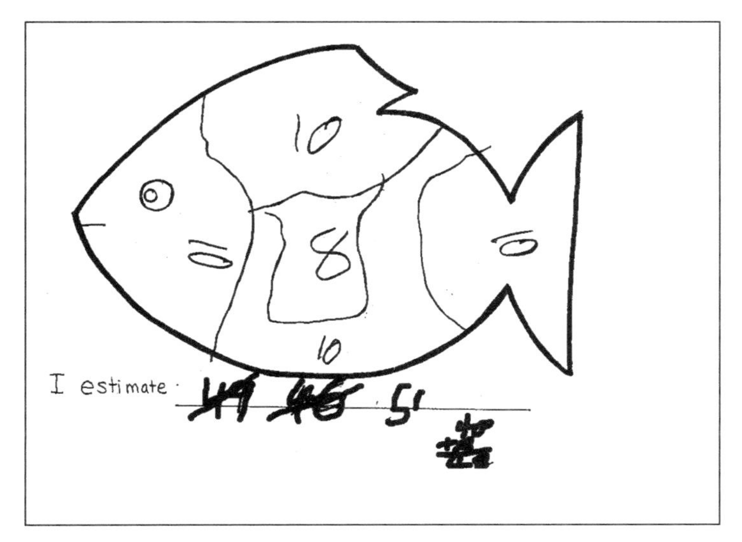

Several students wanted their estimates to be "right." Miles (left) first estimated 49 fish, then changed his estimate to 48. Working at a table with other students who decided that 48 was the "right" answer, Miles ignored his own estimates and spaced the crackers so that he used 48, too.

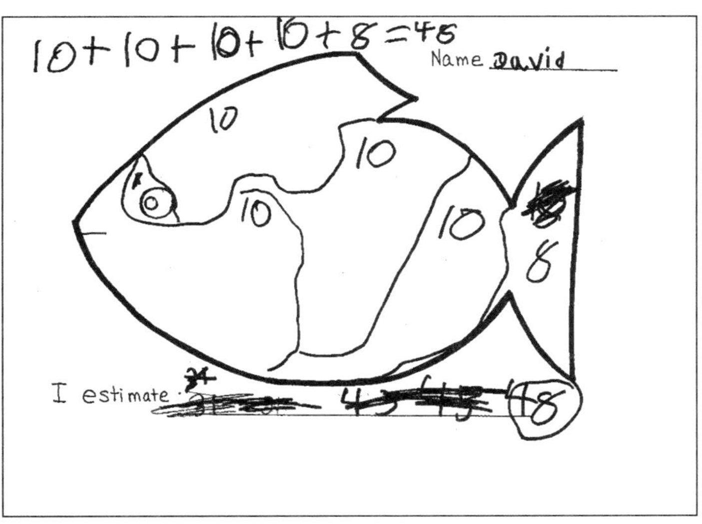

Some students finished early. We asked them to write an addition equation for their fish picture. Many wrote number sentences with multiple addends of 10.

In some cases, we asked students to write another equation that used only two numbers that added up to the total number of fish crackers.

Brian (right) loved testing out different equations. He knew that the order of the addends didn't matter.

Not all of Brian's equations are correct. In an oral interview, Brian corrected himself as he explained how he arrived at each sum.

25
+30
45

35
+10
45

5+5+10+10+10+5=45

10+35=45

10
+35
45

Lesson 9 Make a New Ten

Materials

- **Ten-frame Workmat**, page 110
- **Ten-frame Math 4**, page 115
- **Spinners**, page 111
- **Make a New Ten,** S-8
- crackers (or other counting materials)
- paper clips
- tagboard
- scissors
- pencils
- overhead transparency
- overhead projector

Advance Preparation

Prepare transparencies of a **Ten-frame Workmat**, page 110 and **Ten-frame Math 4**, page 115. Make copies of the **Ten-frame Workmat** on tagboard for each student. You may wish to laminate these for better durability. Make copies of **Make a New Ten,** S-8, for each student.

Copy spinners onto tagboard and cut sheets in half so that each student will have a spinner. Cut out a spinner for yourself to use to demonstrate this activity. To spin the spinner, us a pencil to anchor a paper clip at the center of the circle. Then, spin the paper clip with your finger. (See drawing below.)

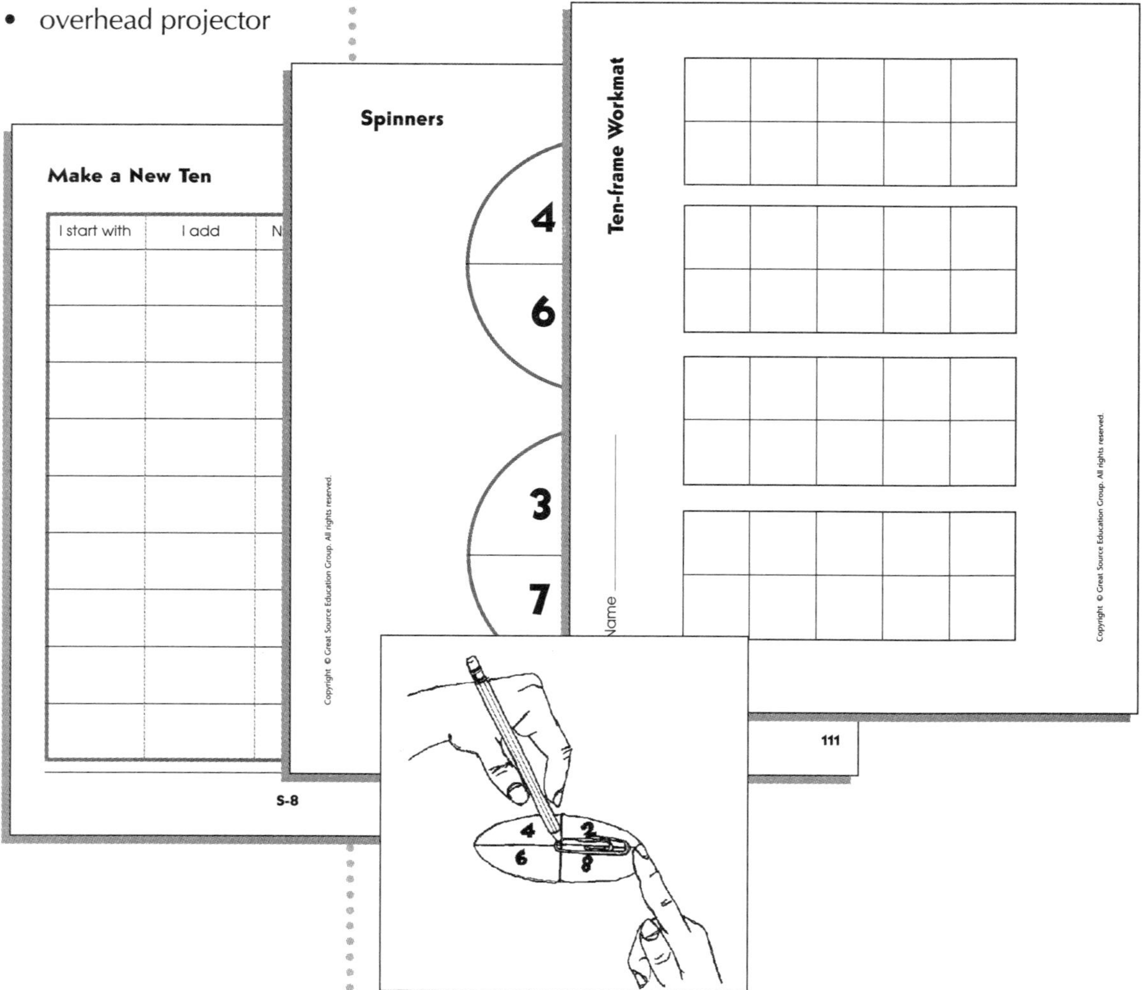

Objectives

- to model two-digit addition using ten-frames
- to add two-digit numbers with and without regrouping

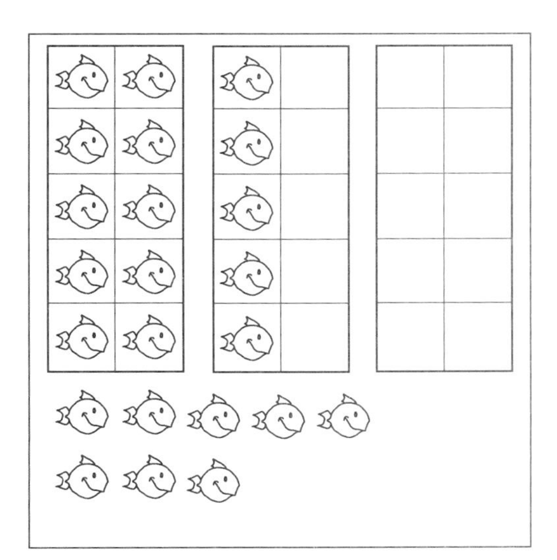

Getting Started

As a warm-up, present **Ten-frame Math 4**, using a transparency of page 115. Then teach the day's lesson.

Display a transparency of a ten-frame workmat on the overhead. Tell a fish story. "There were 15 fish swimming in the ten-frame." Draw 15 fish or place 15 fish crackers on the ten-frame workmat as shown. Continue the story, "Suppose eight more fish swam into the picture. How many fish are there now?" Invite students to share their ideas and how they decided.

Then ask, "When I add eight more, will I be able to complete another ten-frame? How do you know?" Elicit that you could complete one more frame and you would have three fish to begin a third ten-frame.

Then ask, "How many more fish would I need to have 30 fish? How do you know?" Have students share their ideas.

Finally, explain that today students will be playing a spinner game and adding on a ten-frame. Demonstrate **Make a New Ten** (see page 52.)

TEACHING TIP

CLASSROOM MANAGEMENT

When students are organizing for an activity, give them an organizational plan such as, "Go to tables. We need two boys and two girls at each table."

Make a New Ten

Distribute materials for spinners. Have students cut out their spinners. Then demonstrate how to use a pencil and a paper clip to spin a spinner. Give students time to practice using the spinners. Then distribute **Ten-frame Workmat,** recording sheets, and crackers or other counting materials. Have students play "Make a New Ten" (page 52) with partners, in small groups, or as a whole class.

Circulate and observe as students play. This is a good time to have informal interviews with individual students to listen to their thinking as they play.

Wrapping Up

Gather the class for discussion. Ask questions like:

"How did you know when you were going to make a ten?"

"How did the ten-frame help you add?"

Play a Game: "Make a New Ten"

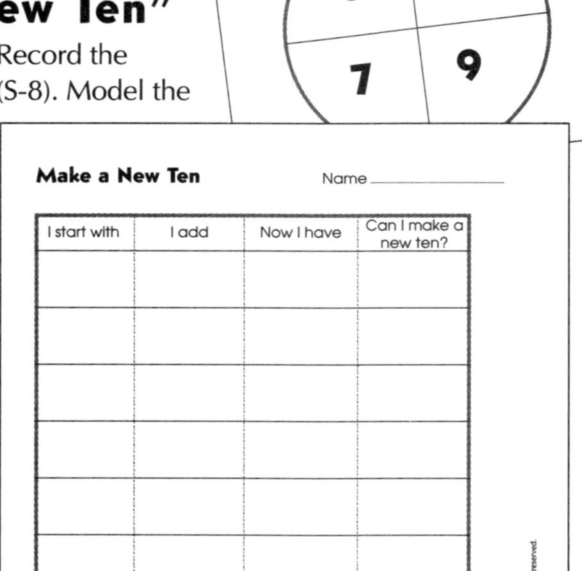

Use your spinner to spin a beginning addend. Record the number in the first box on the recording sheet (S-8). Model the numbers in the ten-frame using fish crackers or other manipulatives. Then have a volunteer spin another addend. Record and model. Use the total from the first round as the first addend of the next equation.

Spin again. Complete the equation. Use the new sum as the first addend of your third equation. Continue in the same way until you reach a total of 40 points. As you play, predict and verify whether you can make a new ten. Write *yes* or *no* in the last column.

Ask questions such as: "Can I fill a ten-frame? How do you know?"

NOTE: There are two different spinners, one with odd numbers and one with even numbers. Students using the odd-number spinner are more likely to reach 40 first. Encourage students to try to figure out why some people get to 40 before others.

What Really Happened

We introduced this activity in small groups of six children. Each group had about thirty-five minutes to play the game and complete the recording. Once students understood how the recording sheet worked, they played again for twenty-minute activity periods.

We began by introducing the recording sheet and playing a demonstration round. Many students immediately saw that they could write in a + and = on the recording sheet to make equations.

We distributed a spinner to each student. Once students started working with a partner, they compared spinners. "Hey, mine has all even numbers!" "Yeah, mine has all odd numbers."

We wanted students to develop spinner skills as a part of this lesson. This was an opportunity to practice some manual dexterity and work out the mechanics of keeping your fingers out of the way of the spinning clip. After spinning a few turns, several students decided it was easier just to pick a number. Then they began to experiment with ways to get to 40 the fastest and ways to make patterns of *yes* and *no* in the last column of the recording sheet.

Sharing Ideas and Strategies

Having the freedom to pick a number rather than to spin a number led to interesting experiments and variations on the game.

3 +	9	= 12	yes
12 +	2	= 14	NO
14 +	6	= 20	yes
20 +	2	= 22	NO
22 +	8	= 30	Yes
30 +	6	= 36	NO
36 +	4	= 40	yes
40 +	4	= 44	NO
44 +	6	= 50	yes

Eren has been fascinated all year with patterns. To guarantee a pattern in this game, she started by writing an alternating pattern of yes *and* no *in the last column. She then decided to choose addends from her spinner rather than to spin random addends. In this way she could match equations to her pattern.*

Eren demonstrated competence composing and decomposing tens to match a pattern. Other students tried Eren's rules but were less successful in correctly predicting the needed addends to make the pattern work.

to get the patrn
yes no yes no uoshly
I was useing 4.
so all I needid was a 6.

Eren used her math journal to describe her variation of the game and her strategy for predicting whether she could make another ten. She wrote, "To get the pattern yes-no-yes-no, usually I was using 4 so all I needed was a 6."

We played this game several days in a row. The first day, many students did not use the ten-frame to model their equations. Because the ten-frame was a new tool and the game was new too, students preferred to count on their fingers. While finger counting is a successful strategy when adding one-digit numbers to two-digit addends it became cumbersome with larger addends. On subsequent days, we used Eren's game rules and ten-frames to develop students' competence with the ten-frame as a tool.

Lesson 10

Make a Fish Collage

Materials

- **Big Fish and Little Fish**, S-9
- **Ten-frame Math 5**, page 116
- overhead transparency
- overhead projector
- tagboard (yellow, green, blue, red, and tan)
- **Pattern Block Shapes**, pages 118–122
- pie tins, 3 per worktable
- paste
- plain white paper
- crayons
- pencils

Advance Preparation

See page 16 for involving families in preparing materials for this activity. Copy pages 118–122 onto tagboard. Use colors that correspond to pattern blocks in your classroom. For this project, prepare the following quantities:

Yellow hexagons	15 per sheet	10 sheets
Green triangles	140 per sheet	2 sheets
Blue diamonds	66 per sheet	3 sheets
Red trapezoids	40 per sheet	4 sheets
Tan diamonds	150 per sheet	2 sheets

You may wish to have parent volunteers precut these shapes.

Using the tagboard pattern blocks, make a sample collage of a large fish (with at least 10 blocks) and a smaller fish. Paste the shapes onto a piece of plain white paper.

Just before the lesson, put handfuls of each shape into pie tins (one or two shapes to a tin) and place three tins on each worktable. Also, have copies of **Big Fish and Little Fish**, S-9, ready for student use.

Make an overhead transparency of **Ten-frame Math 5**, page 116.

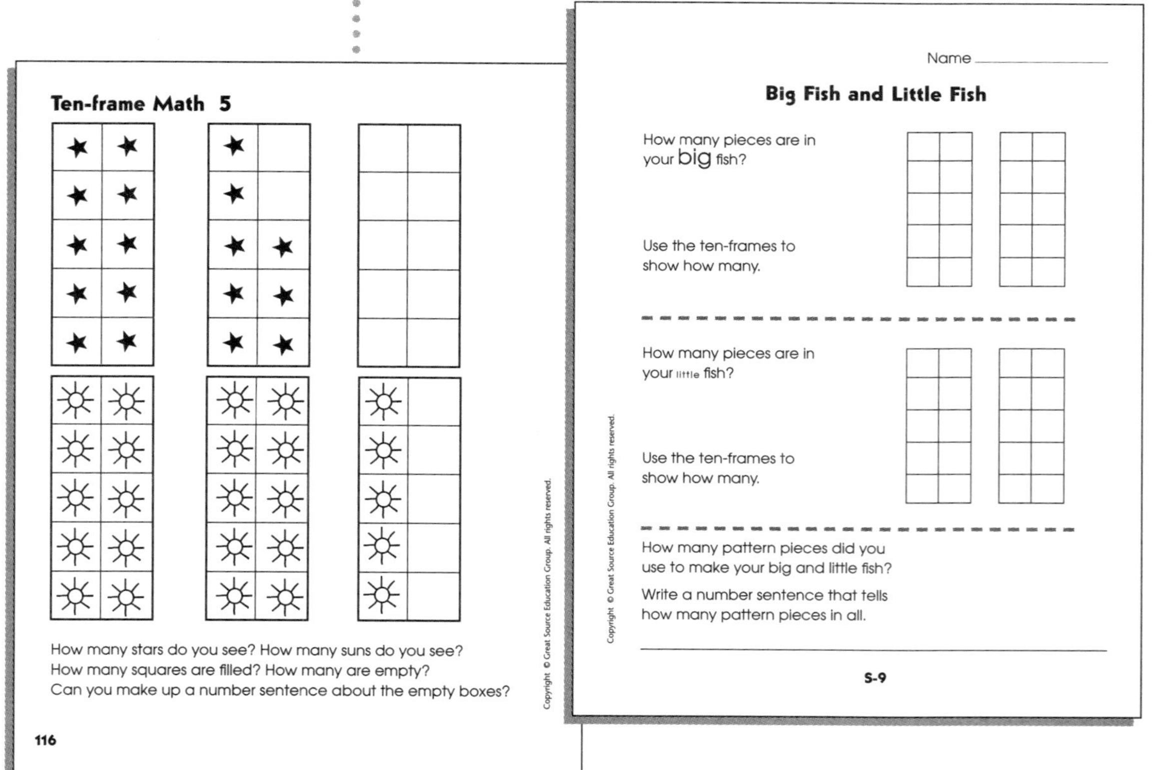

Ten-frame Math 5

How many stars do you see? How many suns do you see?
How many squares are filled? How many are empty?
Can you make up a number sentence about the empty boxes?

116

Name ____________

Big Fish and Little Fish

How many pieces are in your big fish?

Use the ten-frames to show how many.

How many pieces are in your little fish?

Use the ten-frames to show how many.

How many pattern pieces did you use to make your big and little fish?

Write a number sentence that tells how many pattern pieces in all.

____

S-9

Objectives

- to arrange shapes to form a large and a small fish
- to count and classify shapes
- to write an addition equation
- to represent this equation on a ten-frame

Getting Started

As a warm-up, present **Ten-frame Math 5**, using a transparency of page 116. Then teach the day's lesson.

Display the fish collage you have made. Explain that students will create a class aquarium with similar fish collages. Point out that each fish collage will have two fish, a little fish and a big one. The big one should be make up of at least ten pieces. Count the shapes on your sample fish collage.

Explain that first students will play with the shapes, and then create designs. They will paste down their final designs after an adult checks their work.

TEACHING TIP

SUCCESSFUL SEQUENCING IN PASTING

When students paste a design that they have already arranged, have them lift just one or two pieces from the design at a time. Then they won't lose the whole design. If they have overlapping pieces, have them paste the bottom one down first.

Make a Fish Collage

Step 1: Students design their collages. Be flexible as you check the designs, allowing jellyfish, starfish, and sharks as well as traditional fish. However, insist on at least ten shapes for the larger fish. Students will glue their collages.

Step 2: Demonstrate how to write an equation for a ten-frame. Write a prompt on the board:

Big Fish + Little Fish =

Using S-9, have students write about big fish and little fish. Demonstrate how to color in the squares of a ten-frame, one color for the big fish pieces and a different color for the little fish pieces. Count the total number of pieces.

Wrapping Up

Save the collages and ten-frames for upcoming lessons. If students wish, have them draw fish pictures and write fish stories to accompany their collages and equations.

What Really Happened

Eren created complex creatures with many small diamonds. Counting these irregularly-placed shapes proved to be a challenge. She counted 14 shapes in her jellyfish when there were actually 17 shapes, so her graph and her equations did not exactly match her collage.

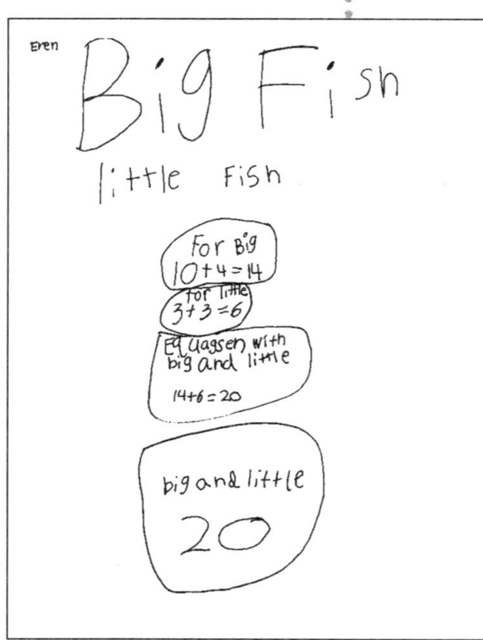

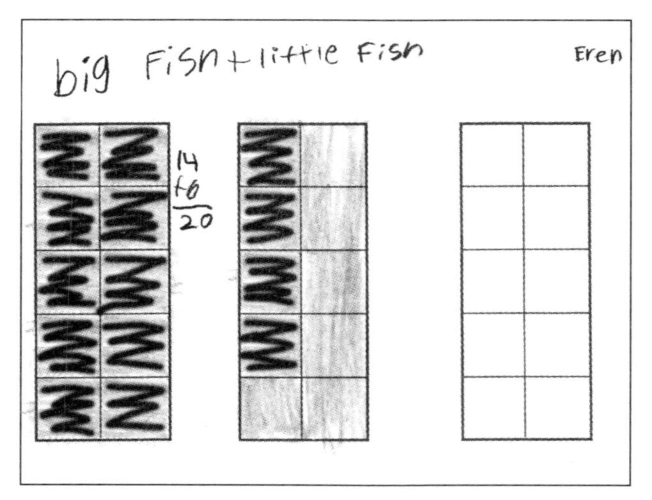

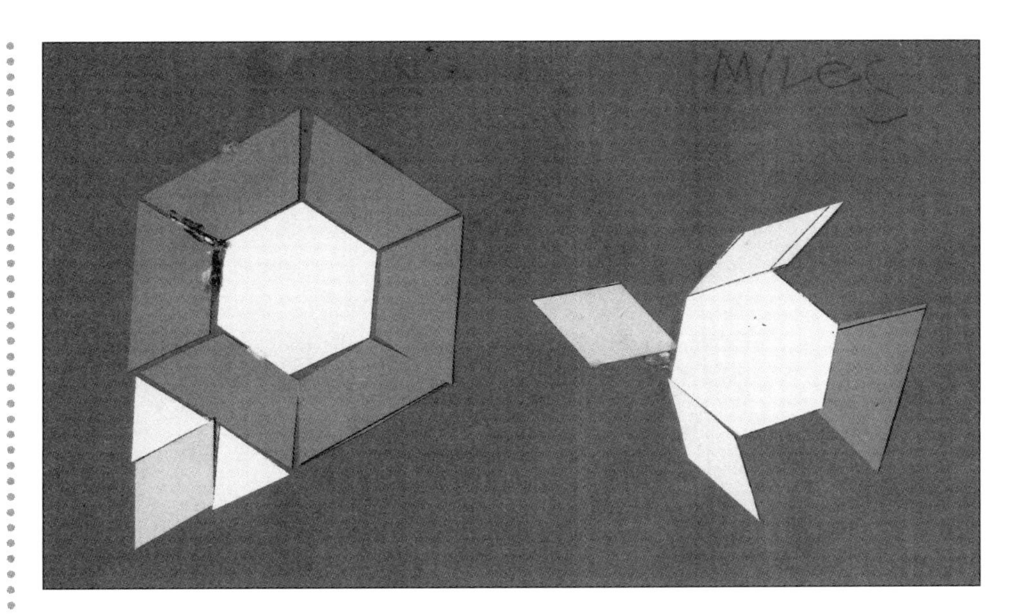

Big fish + Little fish
10 + 5 = 15

Miles' underwater sea creatures both showed symmetry. The pieces for each added up to friendly numbers, 10 and 5, so they were easy to add and chart.

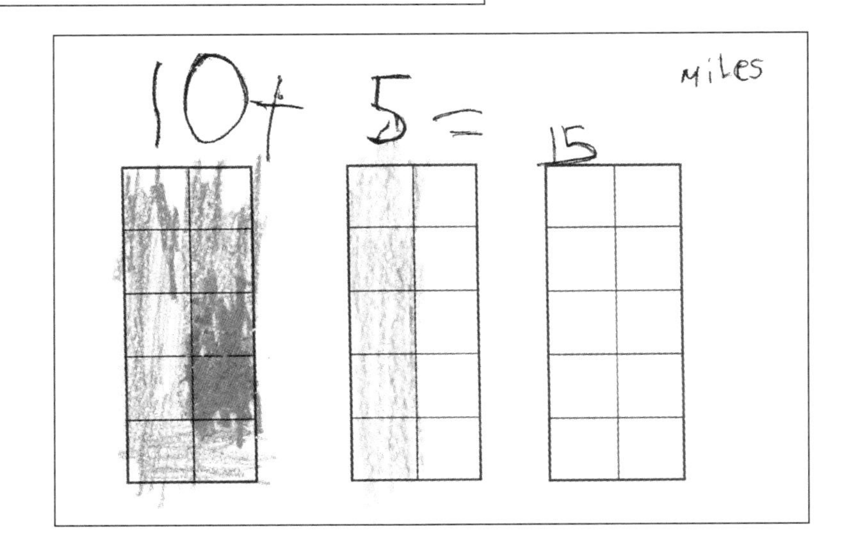

Lesson 11

Count, Add, and Organize Collage Shapes

Materials

- fish collages from previous lesson
- **Ten-frame Workmat**, page 110
- **Ten-frame Math 6**, page 117
- crayons
- pattern blocks that match shapes in collages (or **Pattern Block Shapes**, pages 118-122, made from tagboard)
- small paper bags
- pencils
- overhead transparencies
- overhead projector

Advance Preparation

Prepare transparencies of a **Ten-frame Workmat**, page 110, and **Ten-frame Math 6**, page 117. Also make copies of page 110 for student use.

Gather pattern blocks. If possible, use wooden ones from your classroom set. Or, prepare a duplicate set of tagboard pattern blocks, using reproducibles on pages 118–122. (You will need enough pattern blocks so that all students can match the blocks to the shapes on their fish collages.)

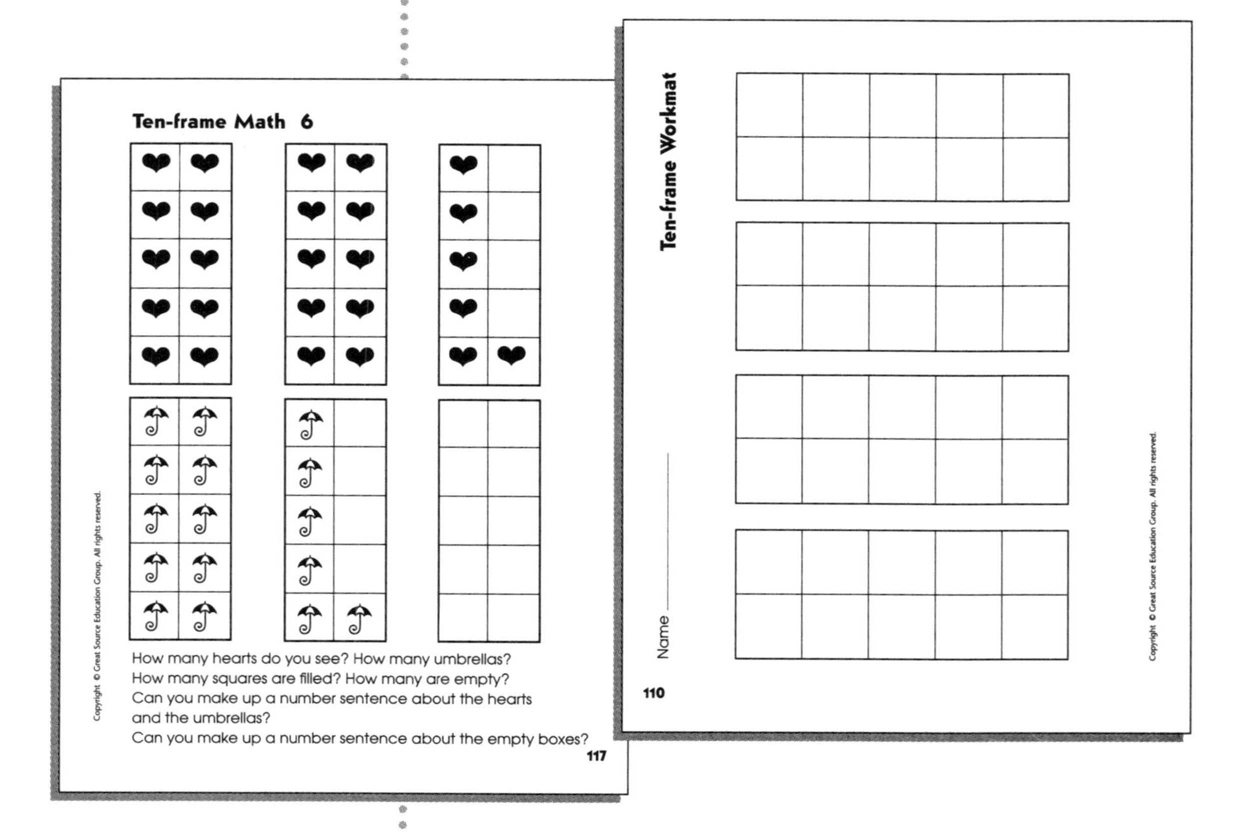

Objectives

- to add two-digit numbers using a ten-frame
- to count and classify geometric shapes
- to organize data before creating a graph

Getting Started

As a warm-up, present **Ten-frame Math 6**, using a transparency of page 117. Then teach the day's lesson.

Display your fish collage. Ask for a volunteer to display his collage. Count the total shapes on each collage.

Then display the transparency of page 110. Color in your total on the top part of the page. (If you have 25 shapes, color two frames plus five extra squares.) Have your partner color in his total on the bottom part of the page. Let the class help you add the tens and the leftovers to come up with a total. Write the equation in the margin.

Juliana carefully traces her shapes onto the paper bag that will store her pattern blocks. She then records how many of each shape she has used in her collage.

Count, Add, and Organize Collage Shapes

Part 1: Have partners work together to find out how many pattern blocks they have used. Record on **Ten-frame Workmat**, page 110.

Part 2: Instruct pairs of students to come to you when they have finished. After you check their work, show them the next project: organizing shapes for an object graph.

Provide pattern blocks that match the collages. Have students place blocks on top of their collage shapes, also counting any overlapping shapes.

Then distribute paper bags. Students will trace a shape on the bag with a pencil; then they count and write the number of that shape on the bag.

Wrapping Up

Students will place their shapes in their bags, ready for the next day's graphing activity.

Lesson 12

Graph Collage Shapes

Materials

- pattern blocks organized in bags from previous lesson
- **Large Block Paper**, 123
- **Make a Graph**, S-10
- **Bar Graph Extension**, 124
- tape
- scissors
- crayons
- pencils

Advance Preparation

Make copies of **Large Block Paper**, page 123, that will serve as a frame for the object graph. If you work in small groups at a table, you will need about 10 sheets in all, as they can be shared and used again. If this is a whole-class activity, estimate the number of sheets you will need.

Make copies of **Make a Graph**, S-10, and **Bar Graph Extension**, page 124, for the students.

Prepare a paper bag with shapes that correspond to your sample collage for the class demonstration.

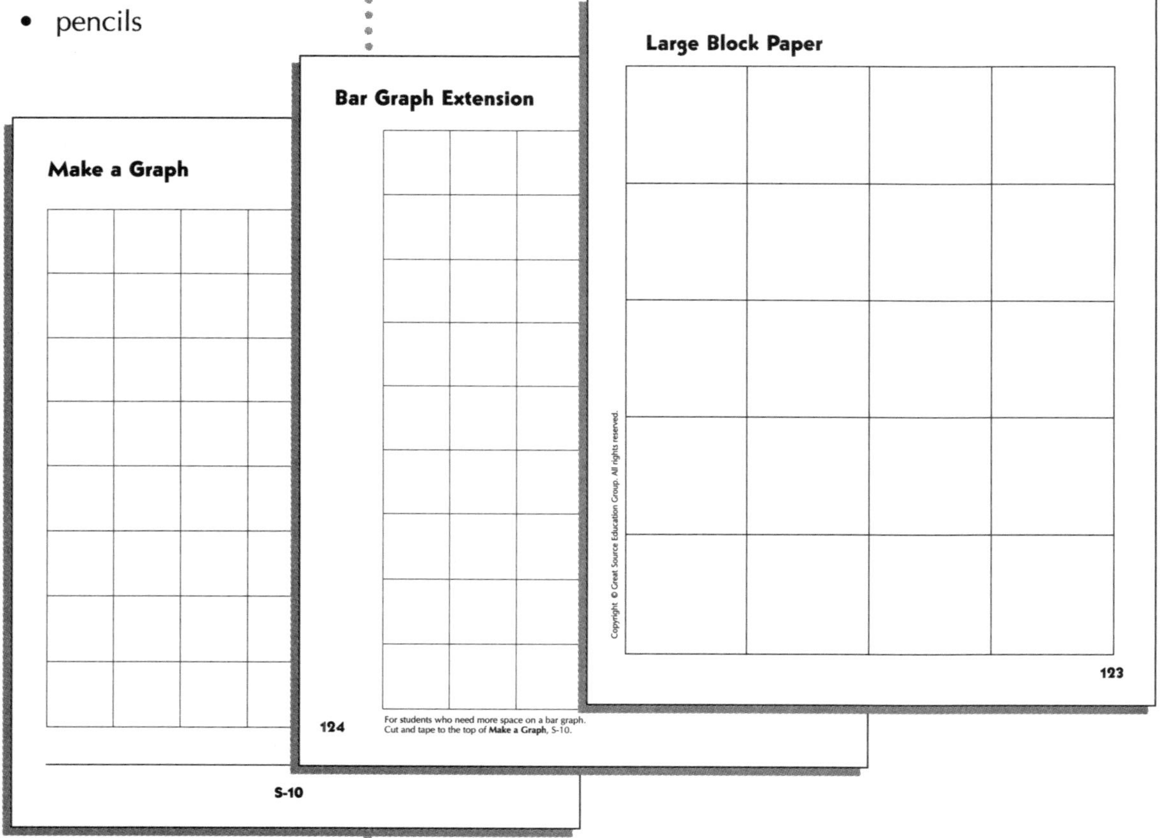

Objectives

- to create an object graph that corresponds to the collage
- to make a bar graph that represents the same data as the object graph
- to write generalizations about graph data

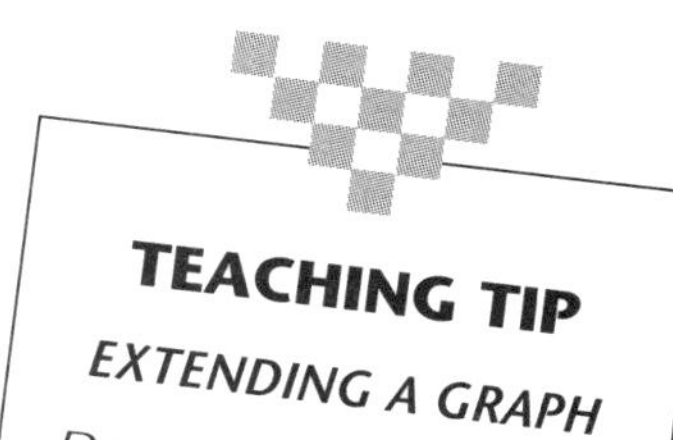

TEACHING TIP

EXTENDING A GRAPH

Demonstrate for students how to tape a copy of page 122 to the top of page S-10. Students can then record large numbers of shapes.

Getting Started

Demonstrate how to make an object graph with your shapes. Display a copy of **Large Block Paper**, page 123, horizontally. Place your shapes on the block paper, with one shape per square. Organize the shapes so that each column contains only one kind of shape. (See below.)

Then display **Make a Graph**, S-10. Count the shapes in the object graph and color, in a matching color, the corresponding squares in the bar graph. Below each bar, draw the shape with which it corresponds. To the left of the graph, number the squares, beginning with *1* beside the bottom square.

Graph Collage Shapes

Have students follow the steps described above using their own bags of pattern blocks.

After students finish their bar graphs, compare the bar graphs and object graphs. Then have students dismantle their object graphs, placing their pattern blocks back into the class supply.

Wrapping Up

Display a few interesting bar graphs and ask what students can tell from the graphs. Compare their observations with the actual collages.

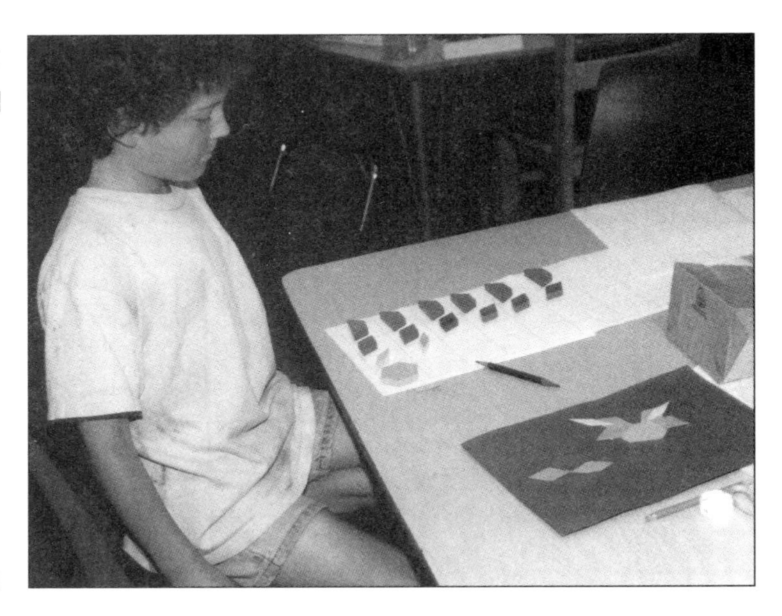

Brian surveys his object graph.

What Really Happened

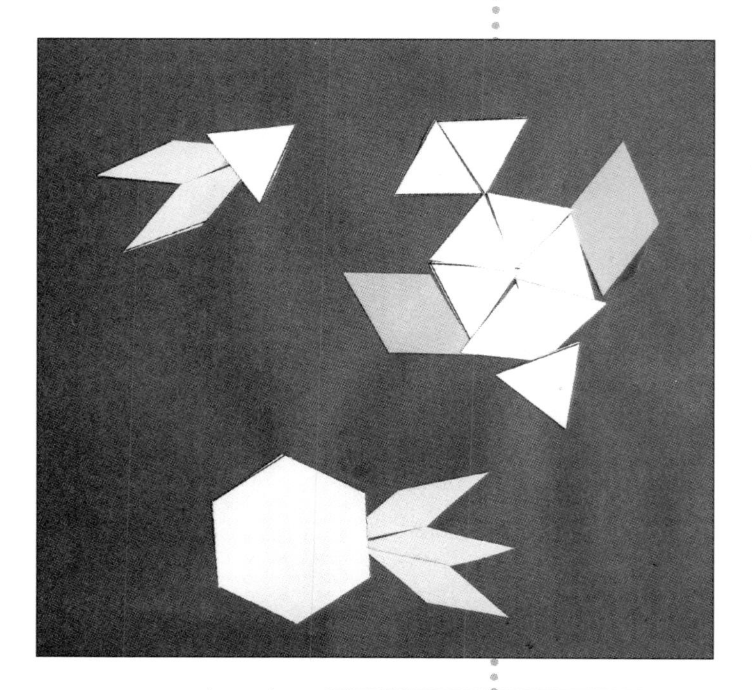

Brennan created three fish so his equation needed three addends. His first equations, though accurate, did not correspond to his picture. When we wondered if he could explain his numbers to us, he scrutinized the picture. Then his face lit up, and he exclaimed, "Now I get it!" and wrote 11 + 4 + 3 = 18.

We gave students the opportunity to write a story or draw a picture about the fish. Brennan enjoyed illustrating his equations page. Allowing students the time to personalize assignments gives them ownership of the project. Notice that Brennan experiments with two ways to write subtraction equations.

Brennan combined numbers as he used a ten-frame to add. He used three crayon colors to distinguish between the counts.

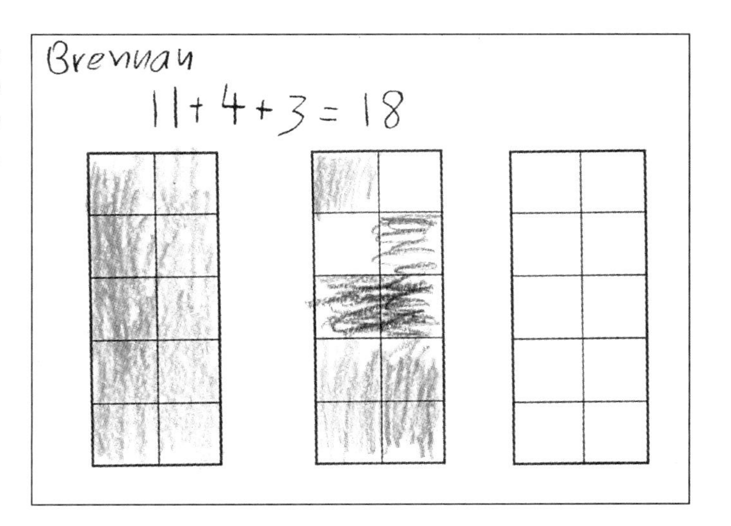

I havd one hexanyon

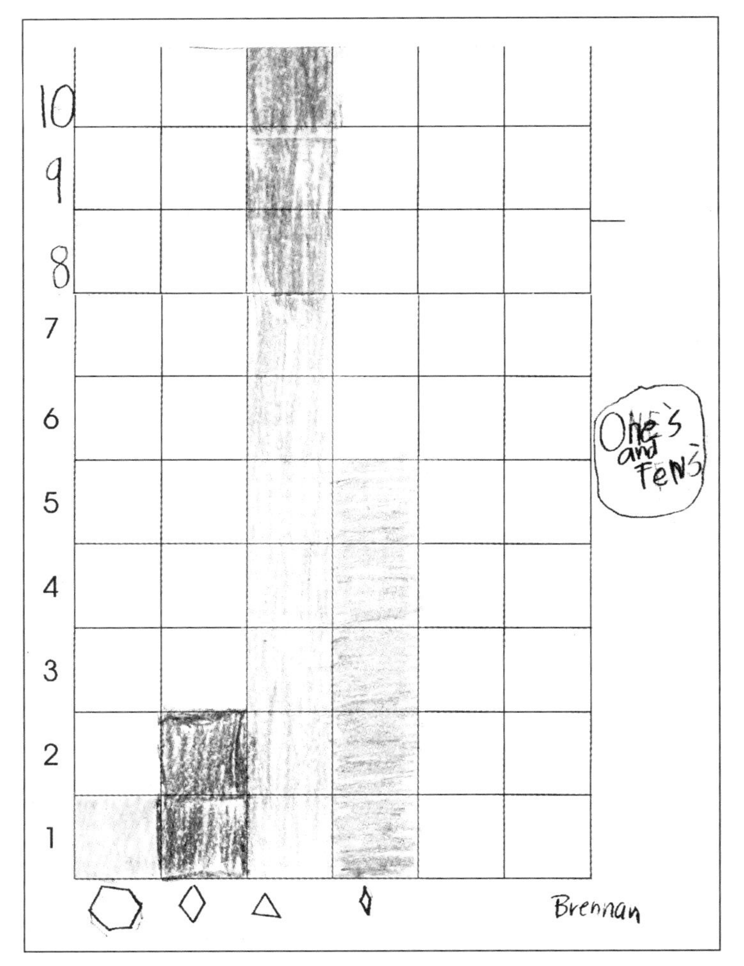

Brennan's factual interpretation of his graph was typical of the class. First graders are very literal and like to use new words.

What Really Happened

Sarah created a large fish out of many bright yellow hexagons. This meant she could work with large numbers, which she enjoyed.

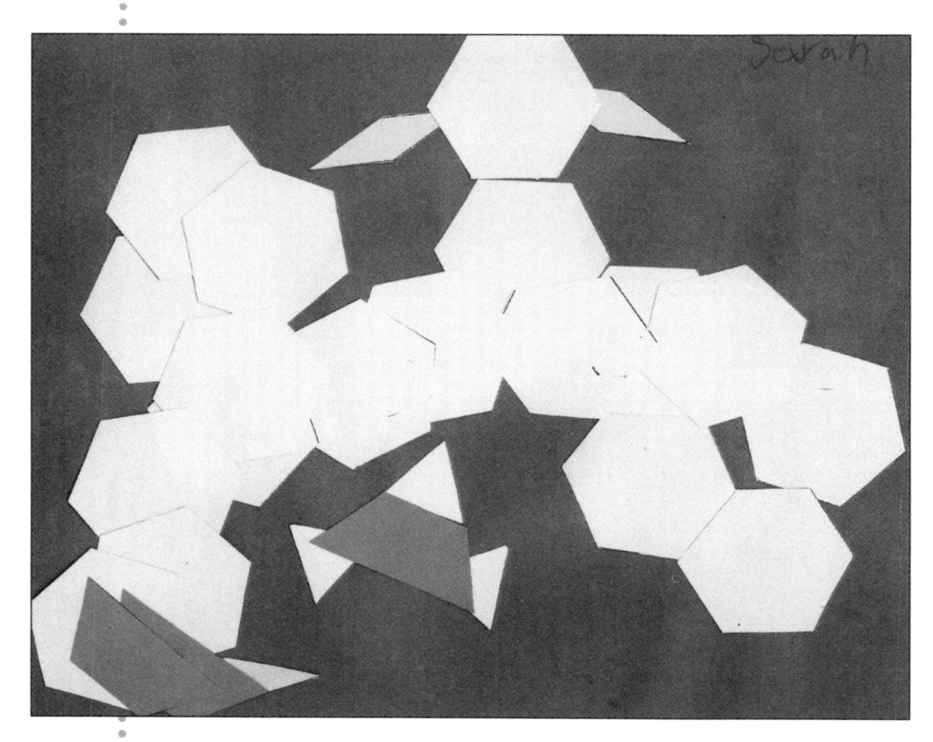

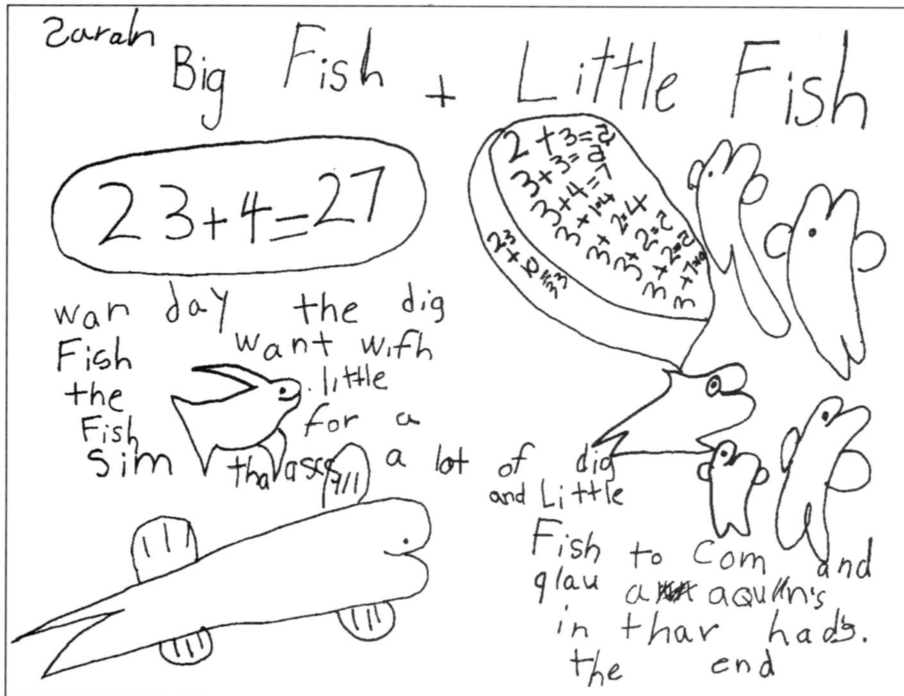

Sarah is a great storyteller. Her story reads, "One day the big fish went with the little fish for a swim. They asked a lot of big and little fish to come and play equations in their heads. The end." An informal assessment of her fish's imagined equations shows that she is still learning addition facts.

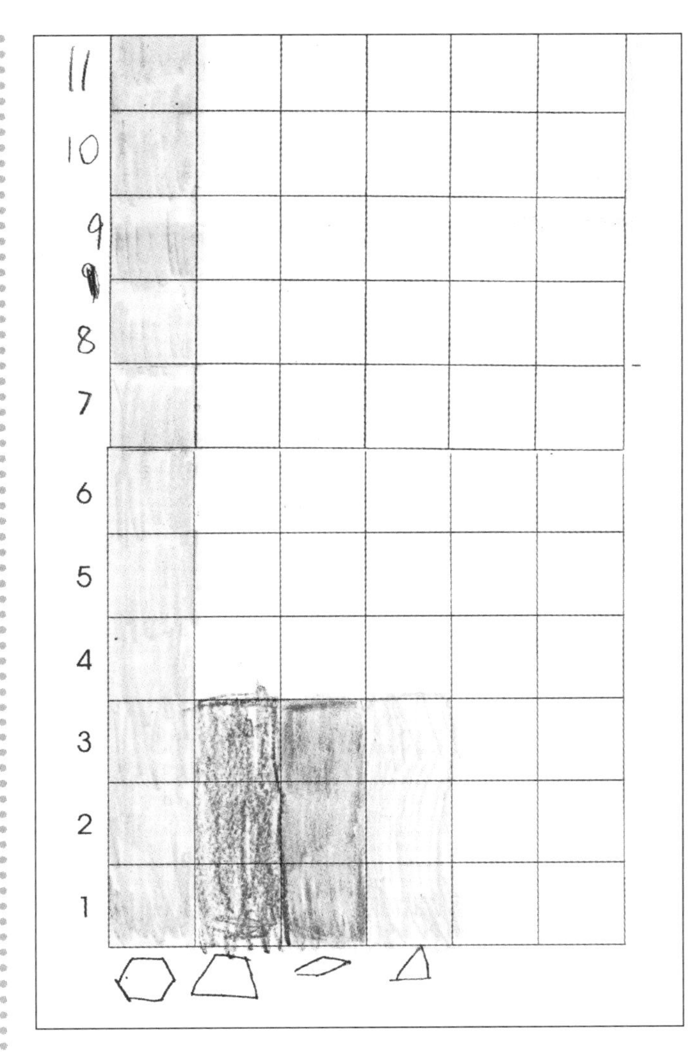

Sarah's object graph with 16 hexagons took up much of the table. We pasted together several pieces of paper for her bar graph so she could record all 16 hexagons. (It is only partially shown here.) She extended the y-axis herself.

Sarah's comments are:
"I have 16 yellow.
I have 3 red.
I have 3 tans.
I have 3 greens.
All of them are 25."

Sarah's ten-frame shows her math. Like Brennan, Sarah added on more than one number within the same frame. She got carried away with her coloring and wrote an equation that did not correspond with her graph.

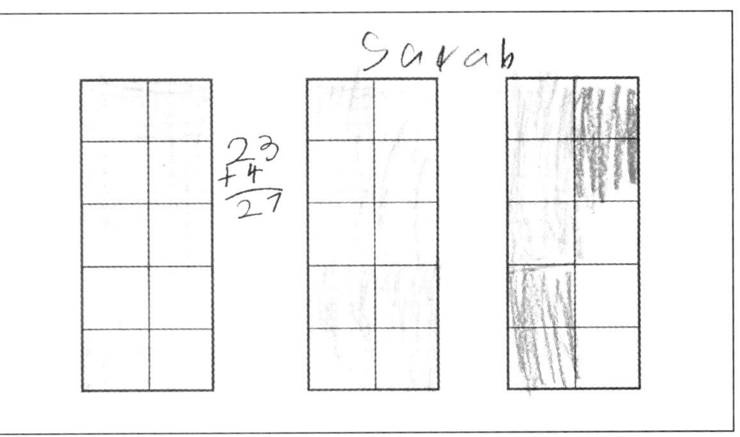

What Really Happened

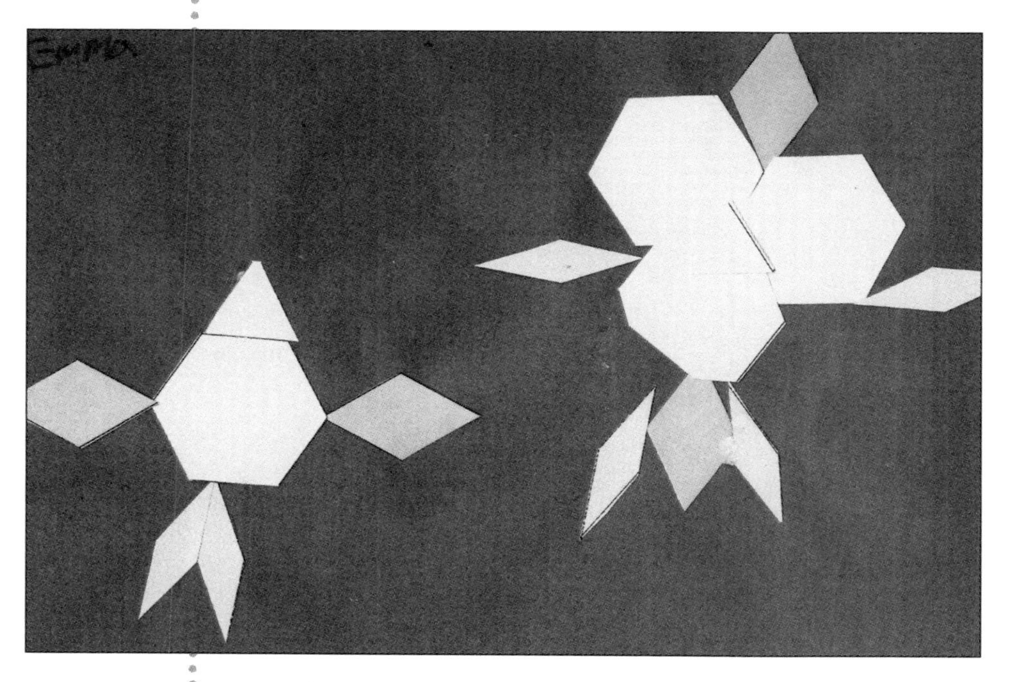

Emma found her fish yielded friendly numbers. She clearly labeled her ten-frame representation and wrote a clear equation.

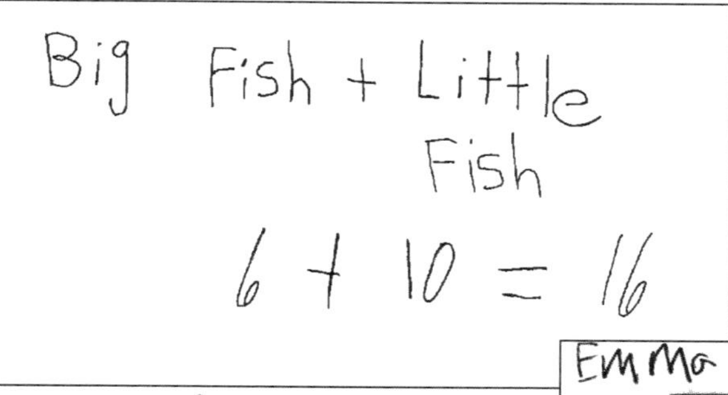

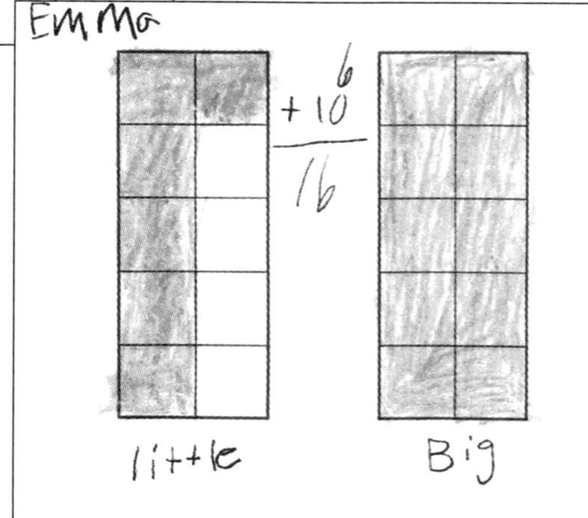

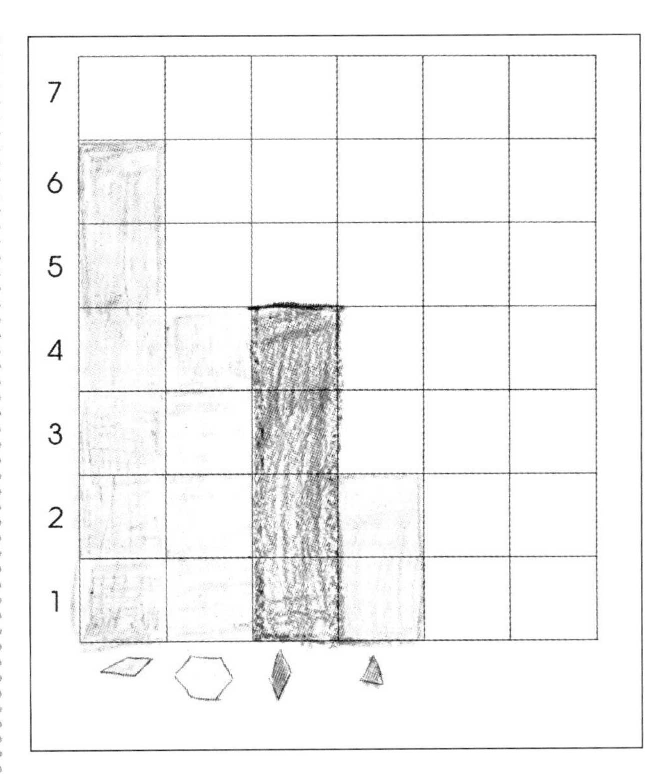

When we asked Emma to write an explanation of her bar graph, like many of her classmates, she protested. She started to write about her two triangles, working with the easiest piece of data. But then she found herself enjoying the project and wrote four sentences, asking us for help with spelling.

there are two triangles.
there are four diamonds.
there are four hexagons
there are six diamonds.

What Really Happened

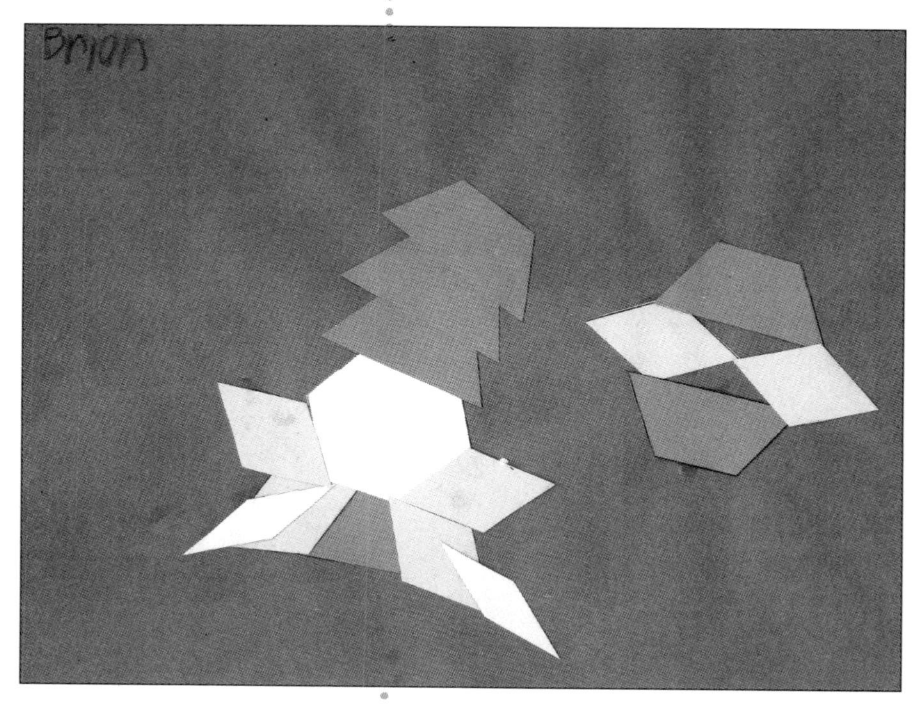

Brian's little fish is imitating the big fish by angling in the water.

Brian's story supports the idea of his picture. "Once upon a time there were two fish. They did everything the other fish did." Brian understood how to write an equation for his picture.

Brian

Big Fish Littl Fish

11+4=15

Onec a Pon a time their Ware tow fish thay did evey thing the other fish did.

the End

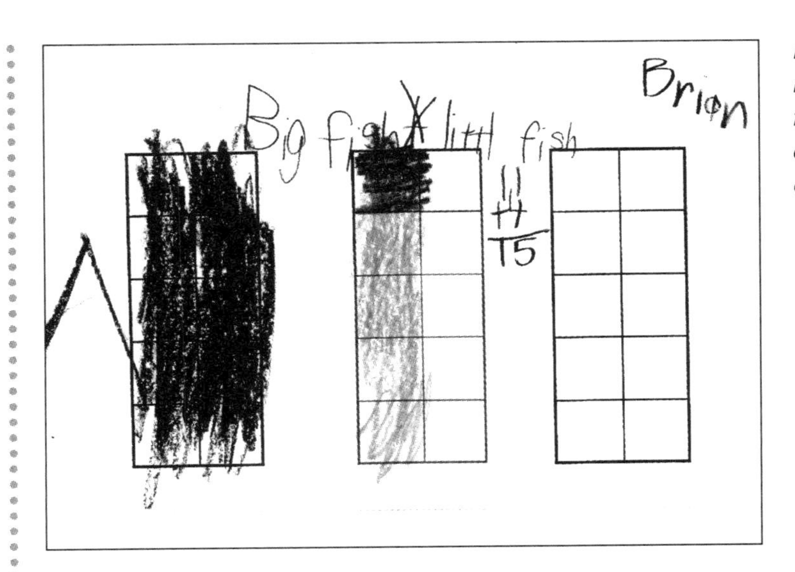

Brian also combined numbers in the ten-frame to illustrate his equation in two colors.

By graphing pattern blocks, Brian discovered that there were the same number of trapezoids as diamonds.

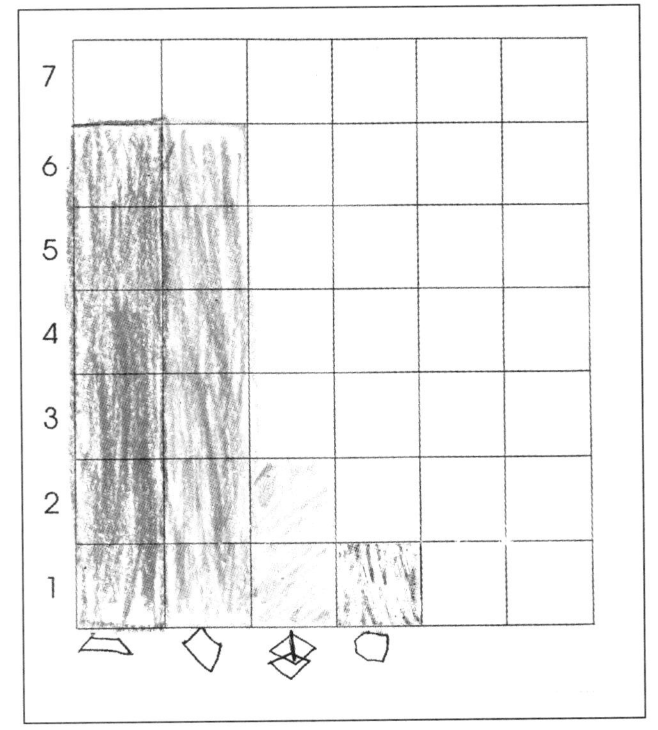

the trapisoyds are the same amownt as the dimind.

Lesson 13 Write and Solve Fish Stories

Materials

- blue construction paper
- fish collages
- **Ten-frame Workmat**, page 110

For **Extensions:**

- **What Does Your Graph Show?**, S-11
- **Tell About Your Graph,** S-12
- **Tell More About Your Graph,** S-13
- colored markers

Advance Preparation

Cover a bulletin board with blue construction paper. Make an aquarium display with students' fish collages. Attach related graphs, stories, and drawings as well.

Have available ten-frames for students to use as they create and solve story problems.

Name ______ Date ______

What Does Your Graph Show?

Look at your fish graph.

How many of each shape did you use?

I used ______ green triangles.

I used ______ blue diamonds.

I used ______ yellow hexagons.

I used ______ red trapezoids.

I used ______ tan diamonds.

Write two sentences about shapes you used.

S-11

Name ______

Tell About Y

Use ten-frames and colored mark

Show how many green triangles you used.

Did you use more green triangles
How many more? How do you kn

S-12

Name ______

Tell More A

Use ten-frames and colored

Show how many yellow hexagons you used.

Did you use more yellow hex
How many more?

S-13

Objectives

- to formulate story problems
- to solve problems involving addition and subtraction
- to assess students understanding of addition and subtraction

Getting Started

Call attention to the aquarium display. (See **Advance Preparation**.) Invite students to read their stories aloud. Choose a graph. Ask, "What does this graph tell us?"

Have students make up an addition story about fish. Solve the story, encouraging students to use ten-frames. Then invite students to make up a subtraction story to solve.

Write and Solve Fish Stories

Have students work independently to write one original addition story and one original subtraction story. Their story problems should be about fish and they may use graphs, ten-frames, or fish pictures for inspiration.

Wrapping Up

Have volunteers present their fish stories for the rest of the class to solve. Discuss and compare solution strategies. Ask questions like,

"How did you solve the story?"

"Did anyone solve it a different way?"

"How many different ways could we solve this story?"

Extensions

Have students write data sheets for their graphs or someone else's graph using **What Does Your Graph Show? Tell About Your Graph** and **Tell More About Your Graph** (S-11, S-12, S-13).

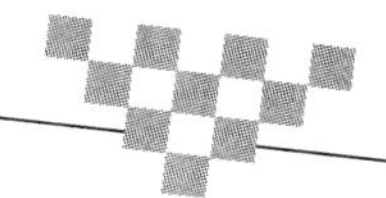

TEACHING TIP

INFORMAL ASSESSMENT

When students write their own word problems, you can assess for their number sense. Do students choose large numbers or small numbers? Does the operation make sense in their story context?

Be sure students explain how they solved their fish stories. Challenge students to prove that their solutions are correct by solving their stories two different ways.

Extensions and Homework

GRAPHING ACTIVITIES:

1. Ask students what types of fish or seafood they like to eat. Help them create a Venn Diagram to show their choices.
2. Invite students to create other object graphs on grids large enough to accommodate such objects. For example, you could group students' shoes in two columns, "Shoes that Tie" and "Other Shoes."
3. Set up a real classroom aquarium and have students study how fish of different species behave. Invite students to record their findings on graphs and in other interesting formats.

MATH ACTIVITIES:

1. Assign S-14 (**Ways to Make Ten**) and S-15 (**More Ways to Make Ten**) as independent work. Call on volunteers to share their results with the class.
2. Have each student choose a number between 6 and 9. Then ask students to make a list of addition facts for their number. Have them compare their list to a friend's.

LANGUAGE ARTS AND ART:

1. Have students write an addition story. Then they trade their story with a classmate and write a number sentence to match the story. Encourage students to draw pictures to illustrate their stories.
2. Tell students to cut out newspaper ads that show different foods or toys. Have them paste the pictures to form a number sentence. Have them tell the addition sentence that describes their picture.

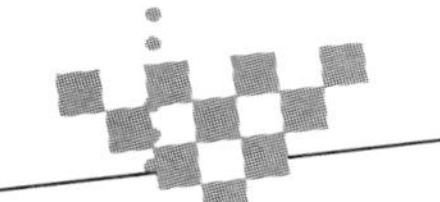

TEACHER NOTE

These are suggestions for extending the lessons in this section.

SECTION 3: STRATEGIES FOR TWO-DIGIT ADDITION AND SUBTRACTION

Lesson 14

Record Marshmallows and Macaroni

Materials

- **Marshmallows and Macaroni,** S-16
- one bag of elbow macaroni
- one bag of miniature marshmallows
- paper cups (one for each student)
- markers in two colors (two for each student)

For **Extension/ Homework:**

- **Writing Equations for Pictures**, S-17

Advance Preparation

Put 12–20 marshmallows and 12–20 macaroni in a cup for each student. Prepare an extra cup for demonstration purposes. Individual students should have slightly different numbers to work with.

Make a copy of S-16 for each student.

First graders may need extra coaching as they begin to write equations for pictures.

Name ____________

Marshmallows and Macaroni

Use ten-frames.
Count your marshmallows.
Record what you see.
Write the number.
____

Use ten-frames.
Count your macaroni.
Record what you see.
Write the number.
____

How many marshmallows and macaroni do you have in all?
Write an equation in the box.

Do you have more macaroni or more marshmallows?
How many more?

S-16

Objectives

- to count and add two-digit numbers using a ten-frame
- to write addition equations
- to explore subtraction as comparing

Getting Started

Explain that today students will be using ten-frames to sort and count marshmallows and macaroni. Demonstrate how to use the ten-frames on S-16 to count the contents of your paper cup.

Review with students what the word *record* means. (It means to write something down.)

Record Marshmallows and Macaroni

Distribute paper cups with marshmallows and macaroni and sheet S-16 to students. Read the instructions on S-16 aloud. If your students are non-readers, you may find it helpful to do the activity as a whole class. If students are working independently, circulate, observe, and interview as needed.

Wrapping Up

Have volunteers share their equations and addition strategies.

Extension/Homework

Have students explore some subtraction situations. For example, say, "Suppose that you eat three of your marshmallows. Now how many do you have left? How do you know? Now eat 5 more marshmallows. How many have you eaten? How many do you have left? How do you know?"

You may wish to assign **Writing Equations for Pictures,** S-17, as homework.

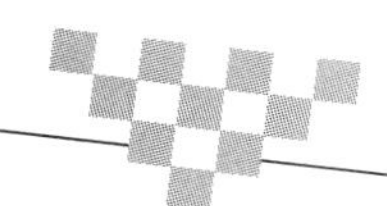

TEACHING TIP

INVOLVING ALL LEARNERS

If you are demonstrating the activity on the overhead, have a volunteer help you use the ten-frame to count marshmallows. Then have another volunteer help you count macaroni. This strategy helps to involve little people who might otherwise be wiggly.

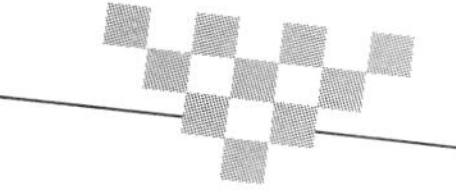

TEACHING TIP

DEVELOPING ORGANIZATIONAL SKILLS

Have students help distribute recording sheets, pencils and cups of marshmallows and macaroni.

What Really Happened

Students were intrigued by this new twist on ten-frame math warmups.

The demonstration cup had 10 marshmallows and 13 macaroni. After we laid them out, students easily saw that altogether there were 23 things in my ten-frames.

Then I asked, "How many more macaroni are there than marshmallows?" Students seemed stumped. I invited them to tell me what they thought I might mean. Using their own words helped them to understand that my question was not as complicated as it sounded.

Use tenframes.
Count your marshmallows.
Record what you see.
Write the number.

10

Use tenframes.
Count your macaroni.
Record what you see.
Write the number.

13

How many marshmallows and macaronis do you have in all?
Write an equation in the box.

$$\begin{array}{r} 10 \\ +13 \\ \hline 23 \end{array}$$

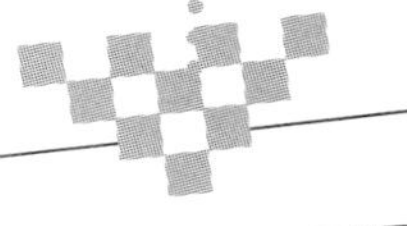

TEACHING TIP

QUESTIONING STRATEGIES

If students find it hard to compare quantities, ask questions like:

"If every macaroni gets a marshmallow, are there enough marshmallows?"

"Why not?"

"How many more marshmallows would you need to give every macaroni a marshmallow?"

"What is the difference between the number of marshmallows and the number of macaroni?"

When we asked students how we could write an equation about marshmallows and macaroni, they offered surprising ideas.

William suggested 2 + 1 = 3. We asked him how he decided on this equation. He said that there are 2 macaroni and 1 more so that makes 3 macaroni.

"What about the other things on the workmat?" we asked.

Matt had an idea. "You have 23 things. Just take away ten and ten more and that leaves 3."

$$\begin{array}{r} 23 \\ -20 \\ \hline 3 \end{array}$$

"Why would this work?" we asked.

"Well, 10 marshmallows cancels out 10 macaroni." Several students agreed with this way of thinking.

"Okay, and how would we write this as a take-away?"

"23 take away 20 is 3."

We demonstrated how to write an equation for his idea on the display copy.

Then we organized students into small groups to explore their own cups of macaroni and marshmallows.

We heard a lot of productive conversation as students discovered how many marshmallows their cups contained.

"Hey, I have 17. Three more and I'd have twenty!"

"I have 19. That's two more than you."

"Christine only had ten. I have 9 more than that."

Sharing Ideas and Strategies

Our students correctly figured out how many macaroni and marshmallows they had altogether. Some even went on to figure out how many two students had altogether. However, most needed coaching by an adult in order to complete all the numerical recording required.

When asked to write their ideas mathematically, most students were able to write several appropriate addition equations. We noticed that students preferred to use horizontal notation. Most students who used vertical notation did not correctly align the numbers but did find correct sums and differences.

Juliana used standard addition as taught to her at home. She saw that she had more marshmallows than macaroni. To find out how many more, she took the total pieces that she had counted, then subtracted the number of marshmallows .

We talked about her equation. "This is a good subtraction equation. What does the answer tell you?"

Juliana looked at it. "It tells me how many are macaroni but it doesn't tell me how many more marshmallows I have." Like many students in the class, Juliana has good instincts about the uses of subtraction but creating a subtraction equation for comparing numbers was complicated.

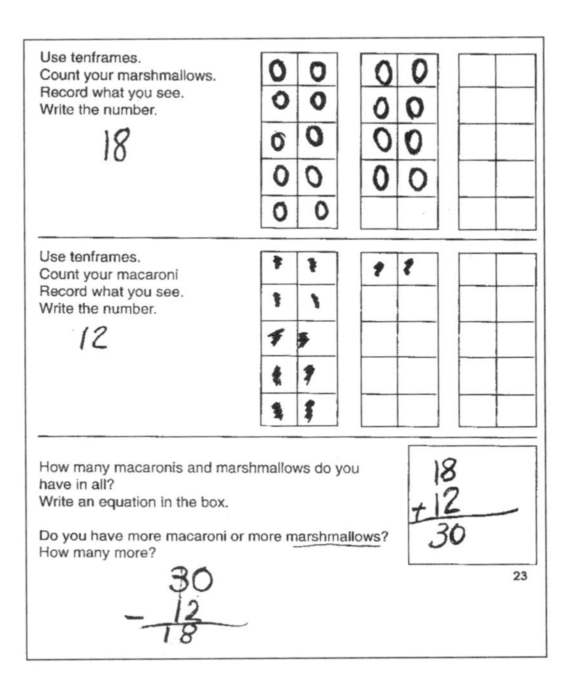
Use tenframes.
Count your marshmallows.
Record what you see.
Write the number.
18

Use tenframes.
Count your macaroni
Record what you see.
Write the number.
12

How many macaronis and marshmallows do you have in all?
Write an equation in the box.
18 + 12 = 30

Do you have more macaroni or more marshmallows?
How many more?
30 − 12 = 18

23

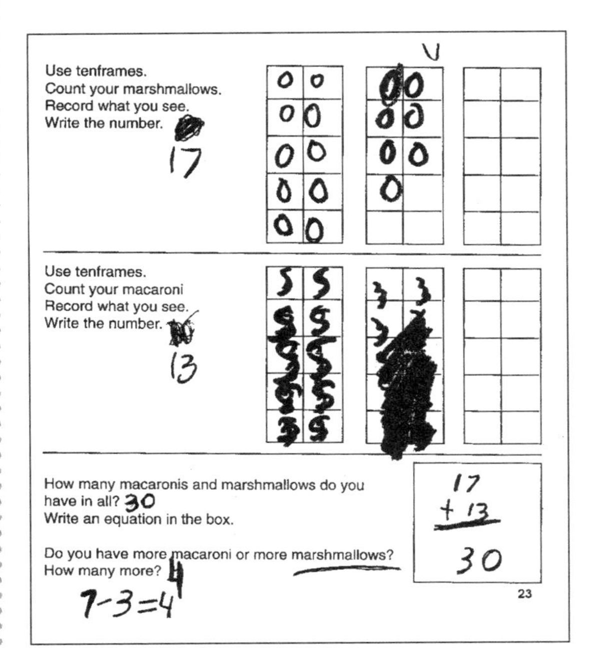
Use tenframes.
Count your marshmallows.
Record what you see.
Write the number.
17

Use tenframes.
Count your macaroni
Record what you see.
Write the number.
13

How many macaronis and marshmallows do you have in all? 30
Write an equation in the box.
17 + 13 = 30

Do you have more macaroni or more marshmallows?
How many more? 4
7-3=4

23

To find out the difference between the number of macaroni and the number of marshmallows, several students first eliminated equivalent sets of tens. Then they wrote subtraction equations to compare the leftovers.

Eren (left) had 17 marshmallows and 13 macaroni. To find the difference, she first eliminated the tens in each set. She knew that she had four more marshmallows than macaroni, so she worked backwards to write a subtraction sentence about the known difference, 7 – 3 = 4.

Lesson 15 TEN-FRAME ADDITION

Materials

- **Writing Equations for Pictures**, S-17
- **Ten-frame Addition,** S-18
- **Show Ten-frame Addition,** S-19
- **Show More Ten-frame Addition,** S-20
- overhead transparency
- overhead projector
- crayons

Advance Preparation

Prepare transparencies of **Writing Equations for Pictures**, S-17 and **Show Ten-frame Addition**, S-19.

Make copies of **Ten-frame Addition,** S-18, **Show Ten-frame Addition,** S-19, and **Show More Ten-frame Addition,** S-20, for student use.

NOTE: When demonstrating how to show ten-frame pictures for equations, help students align the ten-frame recording pages with the related numbered equations as shown below. This strategy will facilitate visual transfer.

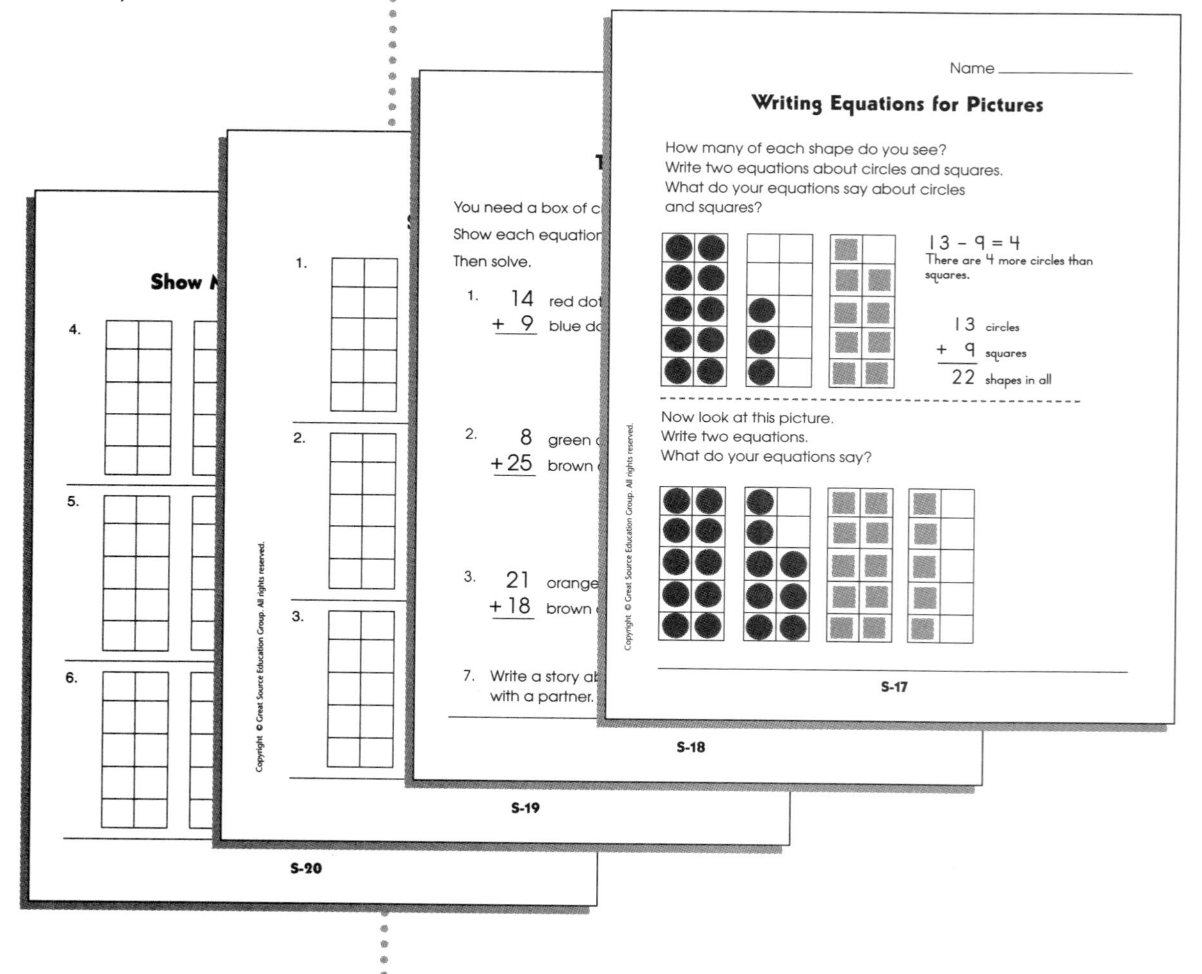

Name ____________

Writing Equations for Pictures

How many of each shape do you see?
Write two equations about circles and squares.
What do your equations say about circles and squares?

13 − 9 = 4
There are 4 more circles than squares.

$$\begin{array}{r} 13 \\ +\ 9 \\ \hline 22 \end{array}$$

13 circles, 9 squares, 22 shapes in all

Now look at this picture.
Write two equations.
What do your equations say?

S-17

You need a box of c
Show each equation
Then solve.

1. 14 red dot
 + 9 blue do

2. 8 green
 +25 brown

3. 21 orange
 +18 brown

7. Write a story ab
 with a partner.

S-18

1.
2.
3.

S-19

Show M

4.
5.
6.

S-20

Objectives

- draw ten-frame pictures for addition stories
- write equations for ten-frame pictures

Getting Started

Use a transparency for S-17, which was assigned in lesson 14, to demonstrate and discuss students' responses.

Then display a transparency of **Show Ten-frame Addition**, S-19. Write a dot story in equation form, for example,

12 black dots
+ 9 orange dots

Have students make up a story about dots to fit the equation.

Ask, "How could we show the story using the ten-frame?" Discuss possibilities. Elicit that students can show black dots and orange dots; then combine dots to show how many in all. Have students demonstrate.

Then ask students to write another equation for the picture. Include both horizontal and vertical forms. If students suggest the commutative pair, 9 orange dots plus 12 black dots, add it to the list of possible correct equations.

Finally, have students make up another dot equation and tell a story about it. Call on volunteers to show how to make a ten-frame picture for the dot story. Have others show two ways to write equations about the picture

TEACHING TIP

MAKING CONNECTIONS

The process of connecting visual and symbolic representations for addition should not be rushed. Make time to interview students about their addition strategies. Invite students to explain to you how they know their answers are right.

Ten-frame Addition

When students are confident with this skill, distribute S-18 through S-20 and have them work independently. Circulate, observe, and interview students about their strategies and their thinking.

Wrapping Up

Have volunteers share their ten-frame pictures. Discuss strategies. Ask questions like:

"How many ways can we write addition equations for a dot story? Let's show all the ways."

Lesson 16 Marshmallow Take-Away

Materials

- small paper cups, one per student
- 20 miniature marshmallows per student
- tagboard
- **Marshmallow Take-Away**, S-21
- **More Marshmallow Take-Away**, S-22

Advance Preparation

Prepare cups of 20 marshmallows ahead of time for fast distribution.

Make copies of **Marshmallow Take-Away**, S-21, and **More Marshmallow Take-Away**, S-22, for student use.

Make number lines by cutting 2 inch strips across lined tagboard. This will create 30 inch strips. Distribute a strip to each student. Guide students to write numbers in sequence, one number at each line, to make their own number lines.

NOTE: This lesson works equally well using ten-frames in place of number lines. Encourage students NOT to eat the manipulatives until they have completed the problem set.

Name ______

More Marshmallow Take-Away

Use a number line and marshmallows.
Write an equation for each story.
Solve each story.

1. 17 marshmallows
   4 rolled away
2. 15 marshm
   9 rolled av
3. 19 marshmallows
   13 rolled away
4. 22 marshm
   18 rolled av
5. 31 marshmallows
   18 rolled away
6. 25 marshm
   6 rolled a
7. I wonder how many marshmallows rolled away in all the stories. How would you figure it out?

______

S-22

Name ______

Marshmallow Take-Away

Use marshmallows and a number line or a yardstick.
Act out each story.
Write how many are left.

1. 12 marshmallows
   10 rolled away
   ____ are left
2. 14 marshmallows
   10 rolled away
   ____ are left
3. 16 marshmallows
   10 rolled away
   ____ are left
4. 18 marshmallows
   10 rolled away
   ____ are left

What patterns do you see in the marshmallow stories?

Make up your own marshmallow story to solve.

______

S-21

Objectives

- to model subtraction problems with concrete objects
- to count back on a number line as a subtraction strategy
- to practice recording subtraction equations

Getting Started

Explain that students will be using marshmallows to explore subtraction. Tell this story: "I saw ten marshmallows on a table. Three rolled away. How many marshmallows are left?" Invite volunteers to act out the story. Then have a volunteer draw a picture for the story. Write the equation 10 – 3=___ on the board as you tell the story again.

Finally, demonstrate the problem and model the solution using a number line or a ten-frame on the overhead.

TEACHING TIP

BUILDING VOCABULARY

When explaining subtraction to beginners, use take away as a verb. Students beginning subtraction will understand "Ten take away three" more easily than "Ten minus three." Later, use the verbs interchangeably.

Marshmallow Take-Away

Distribute materials and S-21 and S-22. Have students use their marshmallows and number lines to model the first story with you.

> *I saw twelve marshmallows sitting on a fence. Ten rolled away. How many marshmallows are left?*

Observe students' modeling strategies. When everyone has a solution, have students share answers and counting strategies. Count together with students if they are making errors by counting too fast.

Continue in the same way with other marshmallow stories. Invite volunteers to make up additional stories for the group to solve.

Wrapping Up

Lead a discussion about using number lines to subtract and add. Ask questions like:

> "How could we use a number line to add?"
>
> "How are adding and subtracting related?"
>
> "How could we use ten-frames to picture take-away stories?"

What Really Happened

All of our first-graders enjoyed this lesson. They found that using a number line with concrete objects was a useful tool for modeling subtraction. Having edible math materials was fun, too. We had students who wanted to experiment with larger numbers use imaginary marshmallows for modeling subtraction from higher numbers. You will want to decide what works best for your particular group. Some students need to physically remove marshmallows in order to "see" the math.

We used both yardsticks and laminated paper number lines with miniature marshmallows. Through trial and error we discovered that this activity worked best when students started with no more than 20 marshmallows. We directed students to place one marshmallow by each number on their number lines, beginning with 1, until they had used up all their marshmallows. Then we asked, "How many marshmallows do you have? How do you know? How many could you eat and still have ten left on your number line? Before you may eat any marshmallows, you must explain how you figured out your answer."

Have students place each subtracted marshmallow away from the number line so that they can visualize the subsets as they subtract. For example, 19 – 5 would look like this:

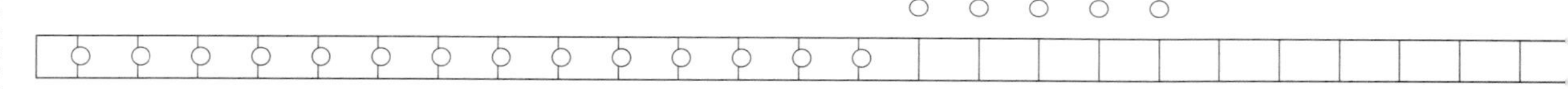

Having marshmallows (or objects) to count allowed students to check their answers by acting out each story.

Connecting take-away actions and subtraction equations was not obvious to all students. When we first asked students to write equations for their own take-away stories, some students wrote addition equations that resulted in the number of marshmallows left on their number line. For example, for a story like "seventeen marshmallows, ten rolled away," students gave equations for the difference, like 3 + 4 = 7. The connections between a modeled action and its mathematical notation were not obvious to many young learners and required many experiences and much discussion.

Students had no trouble reading equations. However, we noticed a wide range of readiness for writing equations. To develop this skill and analyze the connection between actions and equations, we encouraged everyone to explain how their equation matched the action of the story.

Sharing Ideas and Strategies

Students do not naturally align digits when recording subtraction actions vertically. Alignment was not an impediment to correctly solving the problems because students were using equations only as a recording strategy. However, we decided to teach digit alignment at this time, as an important organizational skill.

When modeling equations during discussion, we showed students how to write their one-digit solutions in the ones column, not the tens column. We asked students to write their numbers about the same size as the printed ones so they could better align them.

Brennan wasn't aware of where the 8 belonged in the first equation. In his next try, he lined up the ones and tens better.

$$\begin{array}{r} 19 \\ -11 \\ \hline 8 \end{array} \qquad \begin{array}{r} 20 \\ -1 \\ \hline 19 \end{array}$$

$$\begin{array}{r} 19 \\ -11 \\ \hline 8 \end{array} \qquad \begin{array}{r} 20 \\ -1 \\ \hline 19 \end{array}$$

Robert writes extremely small numbers which make it even harder to align digits. We encouraged him to try writing numbers about the same size as those printed on worksheets.

Extensions and Homework

LANGUAGE ARTS ACTIVITIES:

TEACHER NOTE

These are suggestions for extending the lessons in this section.

1. Retell the story "Jack and the Beanstalk" to the students. Discuss how quickly the beanstalk grew. Then have students work in pairs and make up beanstalk problems that involve adding two-digit numbers. Encourage students to use tens and ones models to solve the problems.

2. Have students make up riddles about two digit numbers, such as the following:

   I am more than 35 and less than 40.

   I have an 8 in the ones column.

   What number am I? (38)

MATH ACTIVITIES:

1. Cut out fish shapes from tagboard. Write a number (1 to 50) on each fish. Attach a paper clip to the "nose" of each fish and place them in a box. Students can take turns lowering a "fishing rod", with a magnet attached to the end, into the box to fish for two fish. Then each student finds the sum or the difference of the two numbers on his or her fish by using models.

2. Prepare index cards with addition and subtraction money problems written on them, such as the following:

   $$\begin{array}{r}.45\\ \underline{+.21}\end{array}\qquad\begin{array}{r}.76\\ \underline{-.33}\end{array}\qquad\begin{array}{r}.32\\ \underline{+.57}\end{array}$$

   Have students work in pairs or small groups and solve the problems using a calculator.

3. Have students play a game called "99." Each group gets a paper bag marked **Tens** that contains several number cards for 1 to 5 and a paper bag marked **Ones** that contains number cards for 0 to 9. Tell students that the first player picks a card from each bag and then subtracts that number from 99. The next player picks a card from each bag and subtracts that number from the difference, and so on. The first group to reach zero wins the game.

ADDING AND SUBTRACTING WITH MONEY

Lesson 17 Money Equivalents

Materials

- "Smart" from ***Where the Sidewalk Ends***, by Shel Silverstein, optional
- play money

Advance Preparation

No advance preparation is needed for this lesson.

Objectives

- to review equivalent values of coins up to $1.00
- to write about mathematics

Getting Started

LITERATURE OPTION: Read "Smart" by Shel Silverstein to the class. Use play money to act out the poem. Ask students what they think of the boy in the poem. Allow time for students to share ideas. Then discuss money concepts. Lead a discussion using questions like the following:

"How much is a penny worth?" (Repeat for other coins.)

"How many cents are in a dollar? a dime? a nickel? a quarter? a half-dollar?"

Money Equivalents

Write About Math: Have students choose a stanza of the poem to write about. Ask each student to describe what the boy in the poem did wrong in the chosen stanza and what would have been a better trade.

Wrapping Up

Gather the class for discussion. Have individual students share their ideas about the child in the poem. Encourage volunteers to model transactions using play money.

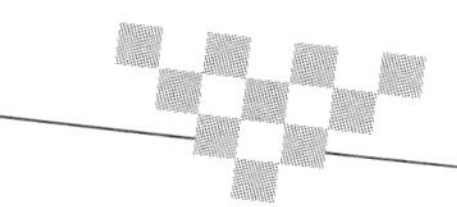

TEACHING TIP

BUILDING VOCABULARY

Sometimes children make wrong calculations because they are unsure of the names associated with coins. Be sure that students have a chance to compare their answers with a partner and discuss the source of discrepancies.

What Really Happened

This lesson was a wonderful success for all students in our combination first/second grade classroom. Everyone easily saw why the child in the poem made terrible mistakes in his exchanges. "When there is money, it is not how many coins you have but what kind of coins you have that count."

This good start created an incentive for all students to explore new and difficult ideas. Some had never worked with money beyond ten cents.

Although they knew their coin equivalences, students had never been asked to show money amounts of more than 10¢. They were very proud of what they learned through this activity.

We brought out play money to act out the exchanges in small groups. Each group acted out the whole poem, recording as each transaction was acted out.

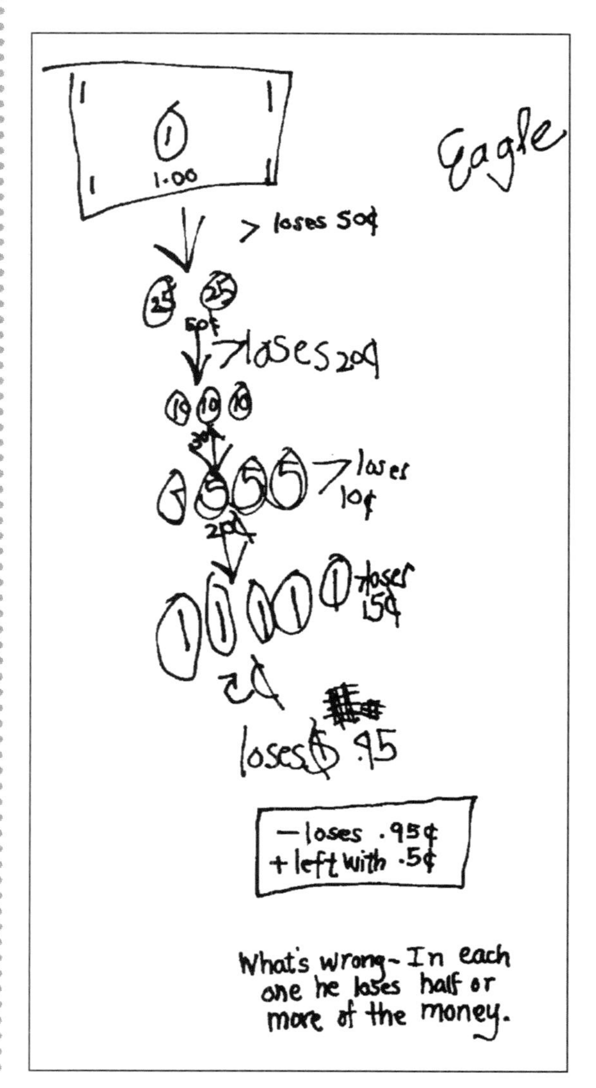

Sharing Ideas and Strategies

Sharing solutions and strategies gave all students a chance to exchange ideas. Students who worked with equations were interested by the pictures their classmates had drawn. Students who had drawn pictures saw the connection between their ideas and the equations and were gratified that they could find a solution in their own way.

Michael

dallar

50¢

30¢

20¢

5¢

hecould Have traded 20 nickls for 1 dallar or one dallar. For fourquarters or onedallar fortendimes.

Students worked through the trades in the whole poem rather than in just one stanza. Several students figured out that they could add up everything the boy lost in each trade, which totaled 95¢. Then they add in the 5¢ he had left over. They know they added right, "because what you lost plus what you have left should equal a dollar."

Lesson 18 COUNT ARTHUR'S MONEY

Materials

- ***Arthur's Funny Money*** by Lillian Hoban
- **Arthur Counts His Money**, S-23
- play money

Advance Preparation

Familiarize yourself with the book ***Arthur's Funny Money***. Prepare copies of **Arthur Counts His Money**, S-23, for student use.

Name ________________

Arthur Counts His Money

Read pages 1-15 in ***Arthur's Funny Money***.

Then answer these questions.

1. How much money did Arthur have in his piggy bank?

$ ____________________

2. Which coins might Arthur have in his bank?

S-23

Objectives

- to solve problems related to money
- to write about mathematics

Getting Started

Read the entire story aloud to the class. Discuss what problem Arthur solved. Have students share experiences they have had saving for something they wanted to buy.

Then explain to students that over the next week or so, they will be studying the money math in the story and will create their own business plans.

Arthur Counts His Money

Distribute copies of S-23 to the students. Read the story problems together. Then have students reread the story as needed to gather the information they need in order to solve the problems.

Instruct students to write their solutions in their math journals or on the reproducible provided.

Allow class time and cooperation between pairs or groups of students to consider the questions and possible solutions.

Wrapping Up

When student groups have finished, have them share their solutions and their strategies. Use play money to model counting money.

What Really Happened

Students discovered that Arthur had a total of $3.78 in his piggy bank. They used money stamps, calculators, and pictures to show the different ways to make $3.78.

First-graders' ability to count this amount of money varied widely. Jenny first drew generic dollars and pennies to show the amount. Then we worked together to show the amount with coin stamps that included dimes, pennies, and half dollars.

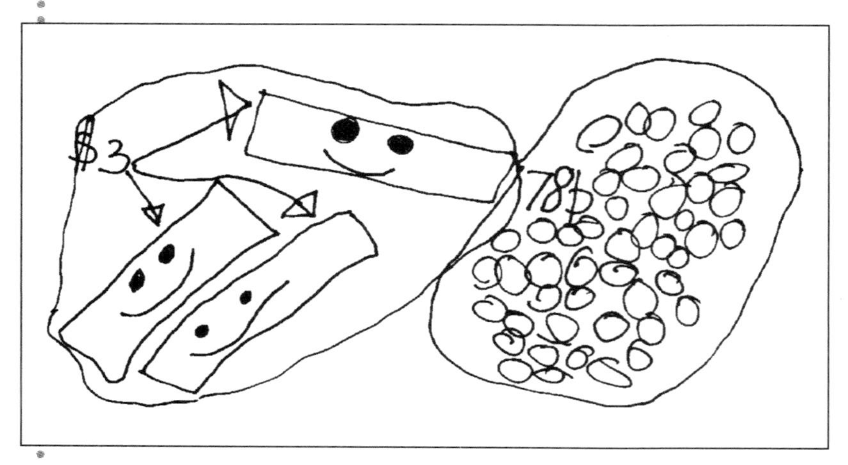

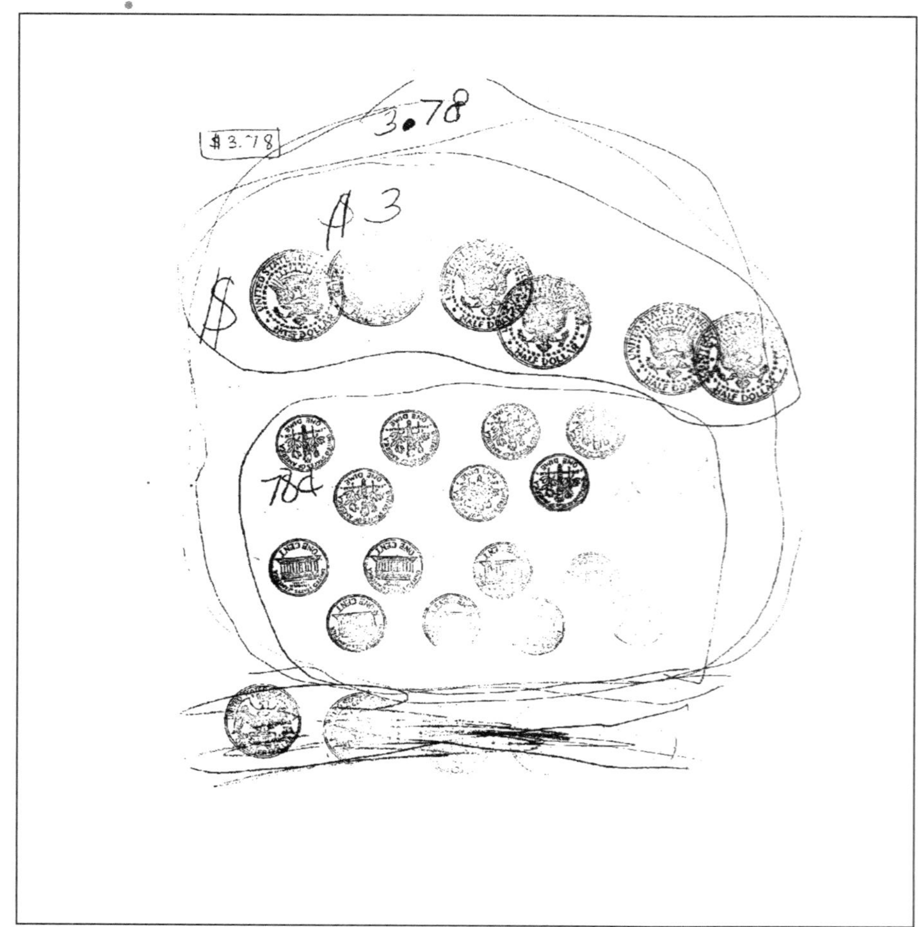

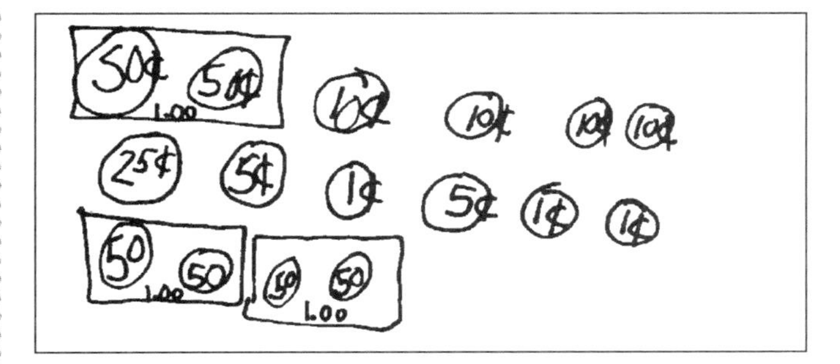

Kathleen grouped half dollars to show three dollars. Then she used combinations of quarters, dimes, nickels and pennies to show one way to make 78¢.

Samantha showed three different ways to make a dollar. Then she showed one way to make 78¢.

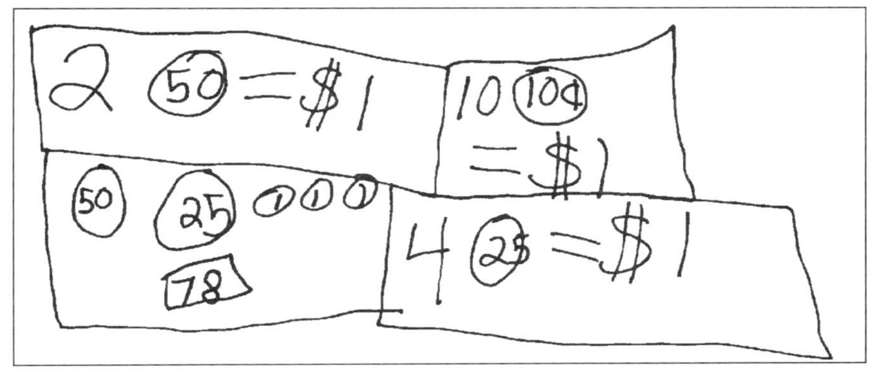

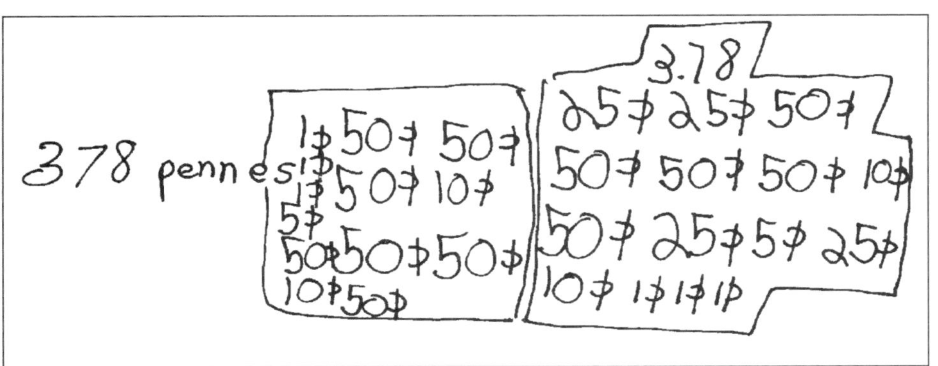

Aaron demonstrated three ways to make $3.78 including with 378 pennies. He didn't feel the need to group his coins for easy counting!

Lesson 19 Open for Business

Materials

- *Arthur's Funny Money* by Lillian Hoban
- **Arthur Sets Up for Business**, S-24
- play money

Advance Preparation

Make copies of **Arthur Sets Up for Business**, S-24, for student use.

Name ____________

Arthur Sets Up for Business

Read pages 16-19 in ***Arthur's Funny Money***.
Then answer these questions.

1. What does Arthur do to set up his business?

2. How much money does he spend on supplies?

3. How much more money will he need to earn?

S-24

Objectives

- to solve problems related to money
- to write about mathematics

Getting Started

Review the previous day's activities. Then have students work independently or with a partner.

Open for Business

Distribute **Arthur Sets Up for Business**, S-24, to the students. Read the questions aloud. Then have students reread the story to gather the information they need to answer the questions.

Instruct students to write their solutions in their math journals. Allow class time and cooperation between pairs or groups of students to consider the questions and possible solutions.

Wrapping Up

When students have finished, gather the students together. Have them share their solutions and their solution strategies.

What Really Happened

We asked our first and second-graders what Arthur needed to do to set up a business. Students picked up on different details. We reread the book with small groups of children who needed extra help.

he had to make a sign.
and he new how mach the
catrmos had to. pay.
he had to pix supplies
80¢
the Soap is. 53¢
the Brillo is. 27¢

Jenny explains that Arthur had to make a sign. He also knew how much the customers had to pay. He had to buy supplies that cost him 80¢.

Samantha drew a picture of the story that shows Arthur, Violet, and the piggy bank. She then wrote about how Arthur planned to set up a business to earn enough money to buy a baseball cap and a tee shirt. Then she wrote about Arthur's dilemma, explaining, "Arthur wanted to buy a frisbee cap and shirt but he didn't have enough money to buy them. [He only had] 3 dollars and 78¢."

Arthur wondid
to by a frisbe cap and shrt
but he didit hav anaf
mane To by them.
3 dolrs and 78¢

Jenny figured out that if Arthur washes 9 bikes he will make \$2.25. She knew that he would earn 50¢ for every two bikes he washed. She then used guess and check with a calculator to figure out how many 25¢ bike washes he needed to earn \$2.25. As part of her thinking strategies, she used a standard addition procedure that she learned at home.

we counted mane be twen \$2.98 and \$5

Alex and Hannah used addition strategies to find out that Arthur needed to earn \$2.02 more to have enough money for his purchases. Alex (above) counted up from \$2.98 to \$5.00. Hannah (below) used several addition equations with easy numbers to find the difference.

\$3.00

\$3.00
+ \$2.98
\$5.98

2.98
+ 2.02
5.00

Keep Track of Arthur's Money

Materials

- ***Arthur's Funny Money*** by Lillian Hoban
- **Keep Track of Arthur's Money**, S-25

For **Extension/ Homework:**

- **Keep Track of Money**, S-26

Advance Preparation

Make copies of **Keep Track of Arthur's Money**, S-25, and **Keep Track of Money**, S-26, for student use.

Name

Keep Track of Money

Finish reading ***Arthur's Funny Money***. Then answer these questions.

1. Arthur had $4.48. How much more money would he need to reach $5.00?
2. Arthur bought a cap and shirt for $4.25. How much less than $5.00 did the cap and shirt cost?
3. Suppose Arthur had 30¢ and licorice twists cost 5¢ each. How many licorice twists could he buy?

Explain how you figured it out.

S-26

Name

Keep Track of Arthur's Money

Read pages 20-44 in ***Arthur's Funny Money***.

Record Arthur's earnings and expenses.

What he did	How much he earned	or	How much he spent
bought soap			53¢
bought Brillo			27¢
washed bike and trike	42¢		

How much money did Arthur earn?

How did you figure it out?

S-25

Objectives

- to solve problems related to money
- to write about mathematics

Getting Started

Review the previous day's activities. Then have students work independently or with a partner.

Keep Track of Arthur's Money

Distribute **Keep Track of Arthur's Money**, S-25, to students. Then have students reread pages 20–44 of the story to gather the information they need to complete the chart and to solve the problem.

Instruct students to write their solutions in their math journals or on S-25.

Allow class time and cooperation between pairs or groups of students to consider the question and possible solutions.

Wrapping Up

When students have finished, have them share their solutions and their solution strategies.

Extension/Homework

Distribute **Keep Track of Money,** S-26, as appropriate. Make time for students to explain their computation strategies.

Lesson 21 START A BUSINESS

Materials

- calculators
- chart paper

Advance Preparation

No advance preparation is needed for this lesson.

Objectives

- to discuss ideas for earning money
- to create and solve addition and subtraction problems
- to share ideas with others

Getting Started

Gather the class for a discussion and ask, "What would you like to save up for?" List ideas on the board. Then ask, "How will you earn the money you need?" Suggest that students think about things that they might sell to raise money or things that they might do for someone to raise money. List students' responses on chart paper.

Then have students pretend that they have their own business. Have them develop a plan for opening a business of their choice. Guide them by asking questions.

"What kind of business would you like to own? Why?"

"What things do you need to open your business?"

"How would you get people to shop in your store?"

Start a Business

Write these questions on the board: *What does your business do? How much do you think you will earn? How will you spend your earnings?*

Have students work independently or with a partner to develop a business plan. Circulate and coach as needed.

Wrapping Up

When students have completed their business plans, have volunteers present ideas. To keep everyone involved, have listeners check the presenter's math with calculators.

What Really Happened

Students had many varied ideas for making money. When planning their own business expenses and earnings, first-and second-graders chose numbers that were friendly and easy for mental math. This activity provided a good informal assessment of students' most comfortable mental math strategies.

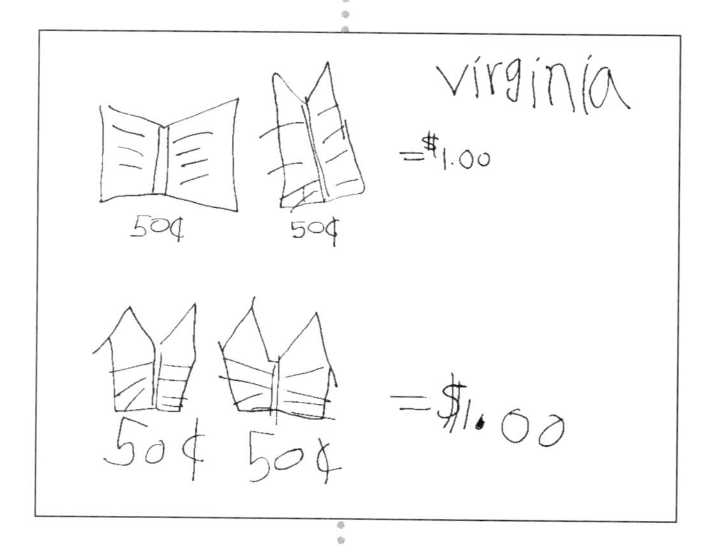

Virginia (left) plans to sell some books. At 50¢ each, she knows she will earn a dollar for every two books she sells.

Stevie (below) plans to wash cars and windows. With the $5.00 she expects to earn, she will buy fish and fish food. She kept track of her expenses and found she would have $1.00 left for some "Jolly Rancher" candies. Her story, like Arthur's, includes some unexpected tasks for which she negotiates an additional fee.

Stevie

I whs a car and I got $4.00 and I made an agrem to wash their windows and I got $1.00 so I had $5.00 and I bot a fish ($3.00) and some fish food ($1.00) and I had $1.00 left so I bot some jlercr.

Chelsea and Jaquie make a list of services and fees. They also accept tips.

wash
Kitchen
take care of Pets
Whs cars for $2.00
take care of Anamells
for $2.00 per Pet -
-fish ,20¢

I exsept tips

They know what they want to buy and exactly how much it costs.

fishing Set for my amarican doll
$27.56
Plus tax's

They use multiplication ideas to figure out how many pets and how many fish they need to care for in order to earn enough for their purchase.

Chelsea
and Jaquie

14 pets / $ 2.00 → $28
3 fish / 20¢ → 60¢

ASSESSMENT 3

Lesson 22 What Have We Learned?

Materials

- **What Did I Learn?**, S-27
- **I Can Solve Story Problems,** S-28
- **Ten-frame Math**, S-29
- **Make a Graph**, S-30
- chart paper
- pattern blocks or shapes (See lesson 10)
- ten-frames
- counters
- play money
- marshmallows
- goldfish crackers

Advance Preparation

Make copies of activity sheets S-27 through S-30 for student use.

You may want to set up an Assessment Center in your classroom with space for up to ten students to complete activity sheets S-28 through S-30. For **Ten-frame Math**, have laminated ten-frames and counters available. For **I Can Solve Story Problems**, have goldfish crackers, marshmallows and play money available. For **Make a Graph**, have pattern blocks available.

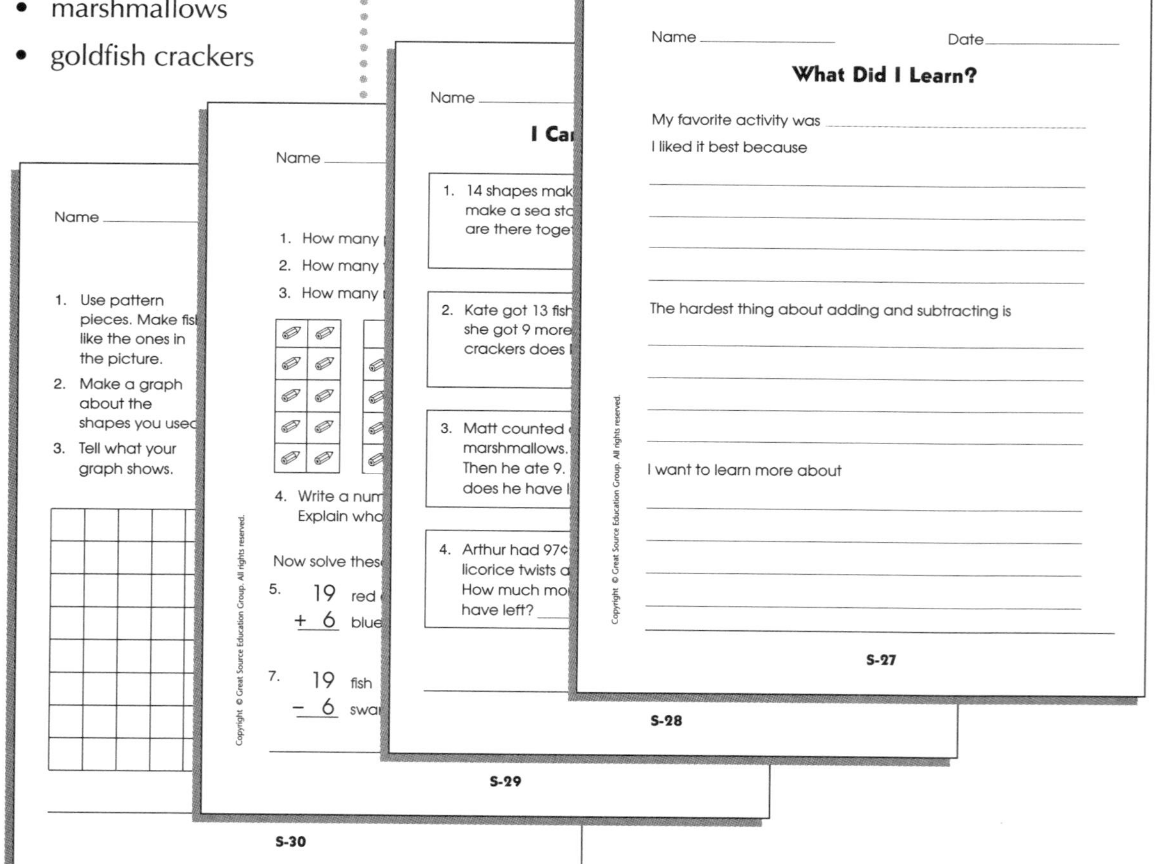

Objectives

- to review addition and subtraction activities
- to reflect on learning
- to assess students' understanding of addition and subtraction

Getting Started

Explain that today students will review what they know about addition and subtraction. Then they will complete different activities over the next few days and write about their learning.

Ask students to help you make a list of the unit activities. Create a chart of their ideas. Then distribute **What Did I Learn?,** S–27. Invite students to refer to the class list of activities as they respond to each question.

What Have We Learned?

If possible, rotate small groups of students through the Assessment Center you have set up. (See **Advance Preparation.**) Each group will work to complete one of the assessments (S-28, S-29, S-30), and then move on to work on another, and so on. As an alternative, plan for the whole class to complete one or two assessments together. Make time to circulate and interview students as they complete each task.

Wrapping Up

Invite students to share their solutions and their problem-solving strategies in class discussions.

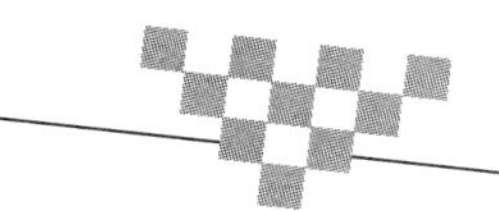

TEACHING TIP

AUTHENTIC ASSESSMENT

Create a supportive evaluation environment. Invite students to look back over the activities they have completed in this unit. For each assessment station, remind them that they have done an activity like this before. Have students think about how each assessment is like other learning they have experienced. Have them tell you how each activity is like previous learning and how it is different.

The assessment items are designed to give a holistic picture of how your students are using their computational skills. While some students may apply standard procedures to solving problems, others may choose to use alternative methods that are equally effective. Be prepared for a variety of solution strategies.

WHAT DID I LEARN?

This activity invites students to review their work and refresh their memories of earlier activities. Some students will have trouble remembering back, so a class list of projects will be helpful here. We find that many students will cite the most recently completed activity as their favorite simply because it is the one they remember best. We honor all answers and ask follow-up questions that let students know we are interested in knowing why they liked a particular activity. "I don't remember the other ones," is a fair and honest answer.

Some students may respond that they didn't like any of the activities. Be sure to honor this opinion as well and ask a follow-up question, for example, "What did you hate most about this unit?" If you hear that activities were boring, ask, "What kind of boring do you mean? Do you mean boring as in too easy or boring as in confusing or boring as in too hard?"

I CAN SOLVE STORY PROBLEMS

Be sure that concrete objects are available as a problem-solving tool. Observe as children work. Interview them about their problem-solving ideas. Have students explain orally or in writing how they solved each problem and how they know their answers are right. This is a good time to apply the communication plan outlined on page 11. You may want to encourage some children to try writing an equation for each story as part of their solution strategy.

TEN-FRAME MATH

Observe as children work. Interview students about their strategies and equations. Some students may write 10 + 4 = 14 pencils. Others may write 14 pencils + 20 flowers = 34 pictures. An unusual equation, but perfectly valid, might be 10 flowers - 4 pencils = 6 more flowers than pencils.

MAKE A GRAPH

The fish can be graphed in different ways. Some children may look at the overall shape or size of each fish and find that there are 3 hexagon (big) fish and 3 diamond (little) fish. Other children may look at the pattern block shapes in the collage and make a graph of the shapes that make up the fish. Yet another solution would be to sort the fish by the number of pattern blocks in each. Be ready for a variety of responses.

Possible comments about what different graphs can show: "There are the same number of big fish and little fish, three each." "You need 20 pattern blocks to make all these fish." "The shape you need the most of are diamond shapes."

Encourage children to try writing equations for the fish picture or for the data in their graph.

TEACHER RESOURCES

Evaluation Checklist

Mathematics Skills	Student											
Knows sums to ten												
Knows differences from ten												
Can add tens and ones												
Correctly compares and orders numbers from one to ten												
Correctly compares and orders numbers between zero and 40												
Predicts outcomes of counting game												
Identifies number patterns in different contexts												
Knows coin equivalents												
Knows skip counting patterns												
Uses skip counting patterns to add and subtract												
Can write meaningful equations for number stories												
Can write meaningful word problems for numbers and equations												

Evidence Key

1. Skip Counting activities
2. Hundred Chart Puzzles
3. Keep the Difference
4. Ten-frame Math
5. Estimating Fish Area
6. Make a New Ten
7. Count and Organize Collage Shapes
8. Writing and Solving Fish Stories
9. Marshmallow Addition
10. Marshmallow Take-Away
11. Money Equivalents
12. Problem Solving with ***Arthur's Funny Money***
13. Assessment 2

Evaluation Checklist

Student											
Strategy Development											
Is comfortable using manipulatives											
Creates/uses visual or pictorial models											
Uses mathematical notation in meaningful ways											
Uses number patterns in meaningful ways											
Can explain decimal sums (money)											
Consciously builds on prior knowledge											
Chooses to use basic facts, mental math, estimation											
Communication Skills											
Asks relevant questions											
Participates in discussion											
Takes turns											
When playing game, follows rules											
Explains rules to others											
Writes and draws clear explanations											

Evidence Key

1. Skip Counting activities
2. Hundred Chart Puzzles
3. Keep the Difference
4. Ten-frame Math
5. Estimating Fish Area
6. Make a New Ten
7. Count and Organize Collage Shapes
8. Writing and Solving Fish Stories
9. Marshmallow Addition
10. Marshmallow Take-Away
11. Money Equivalents
12. Problem Solving with ***Arthur's Funny Money***
13. Assessment 2

Name ______________________

Ten-frame Workmat

Spinners

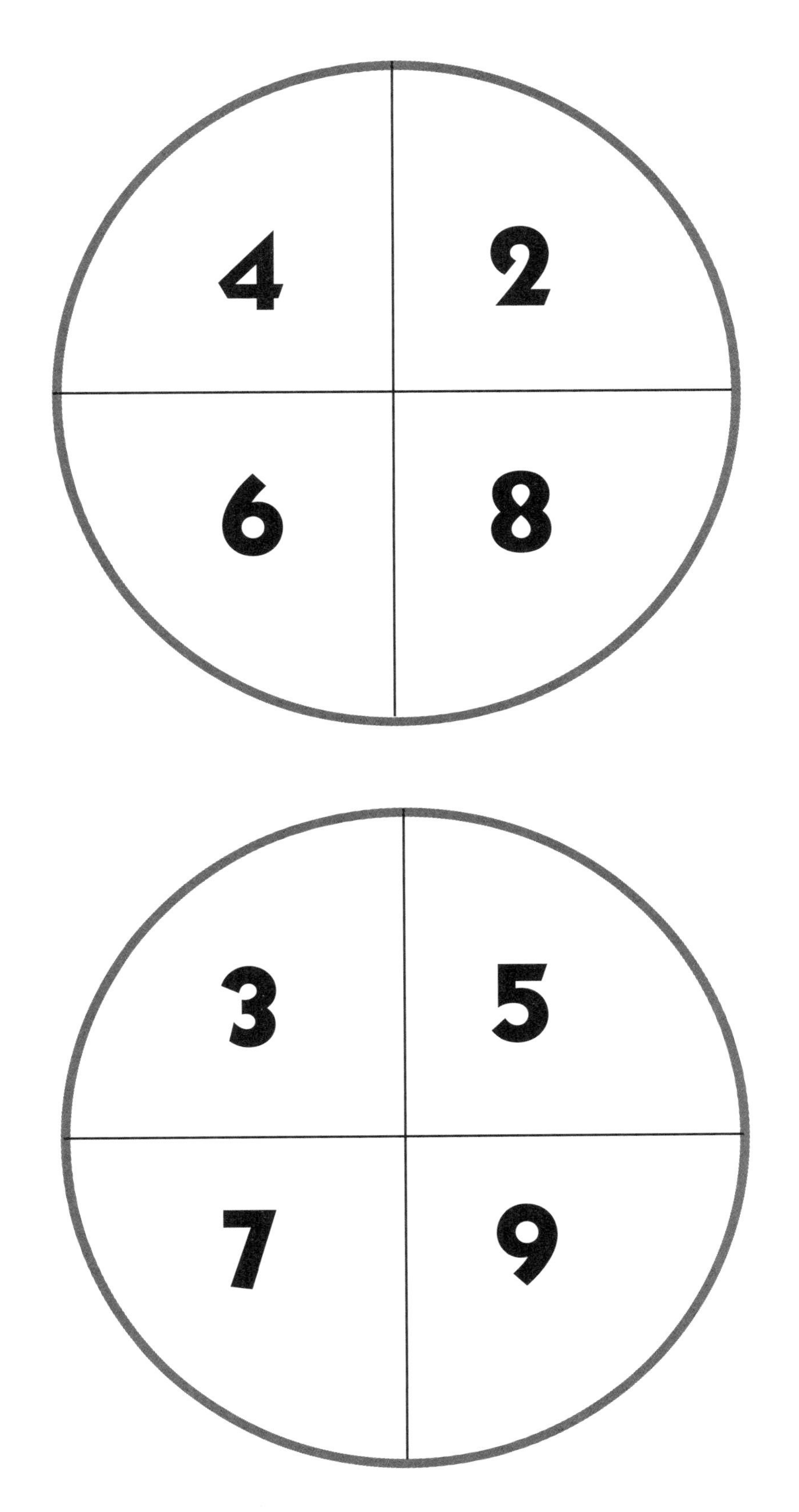

Ten-frame Math 1

How many hearts do you see? How did you figure it out?
How many umbrellas do you see? How did you figure it out?
How many squares are filled? How many squares are empty?

Ten-frame Math 2

How many hearts do you see? How did you figure it out?
How many umbrellas do you see? How did you figure it out?
How many squares are filled? How many squares are empty?

Ten-frame Math 3

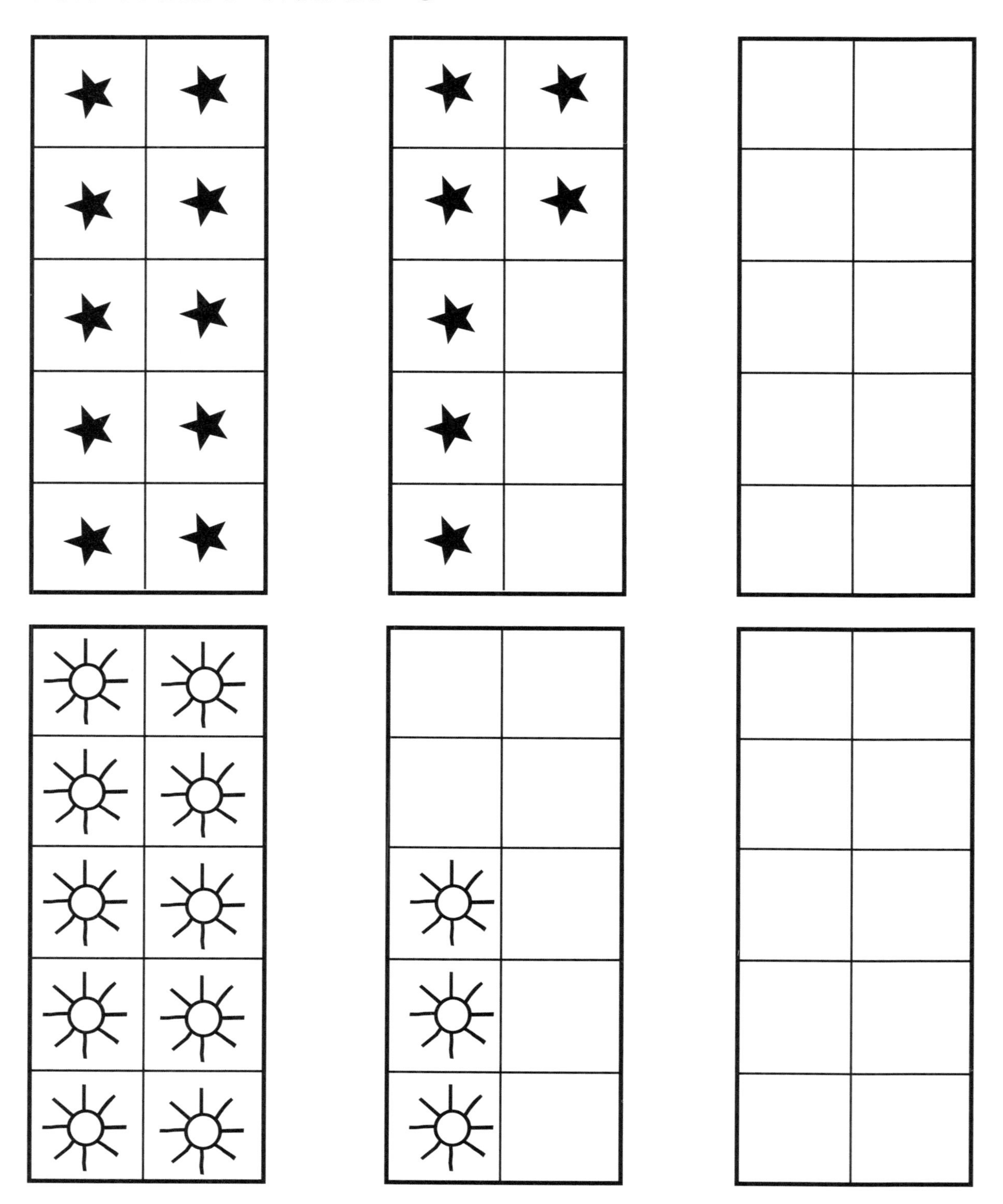

How many stars do you see? How many suns do you see? How many squares are filled? How many squares are empty? Can you make up a number sentence about the stars and the suns?

Ten-frame Math 4

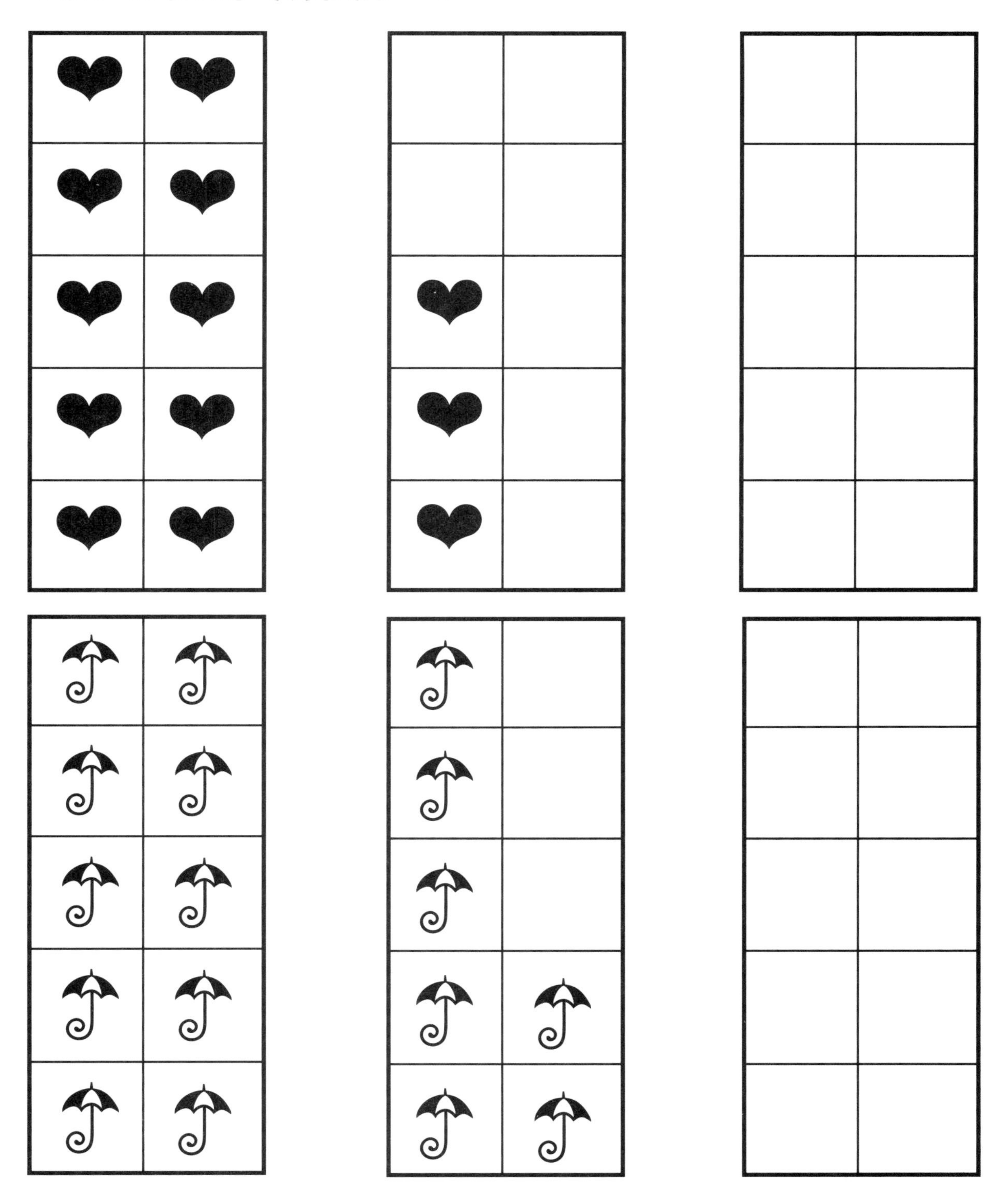

How many hearts do you see? How did you figure it out?
How many umbrellas do you see?
How many squares are filled? How many squares are empty?

Ten-frame Math 5

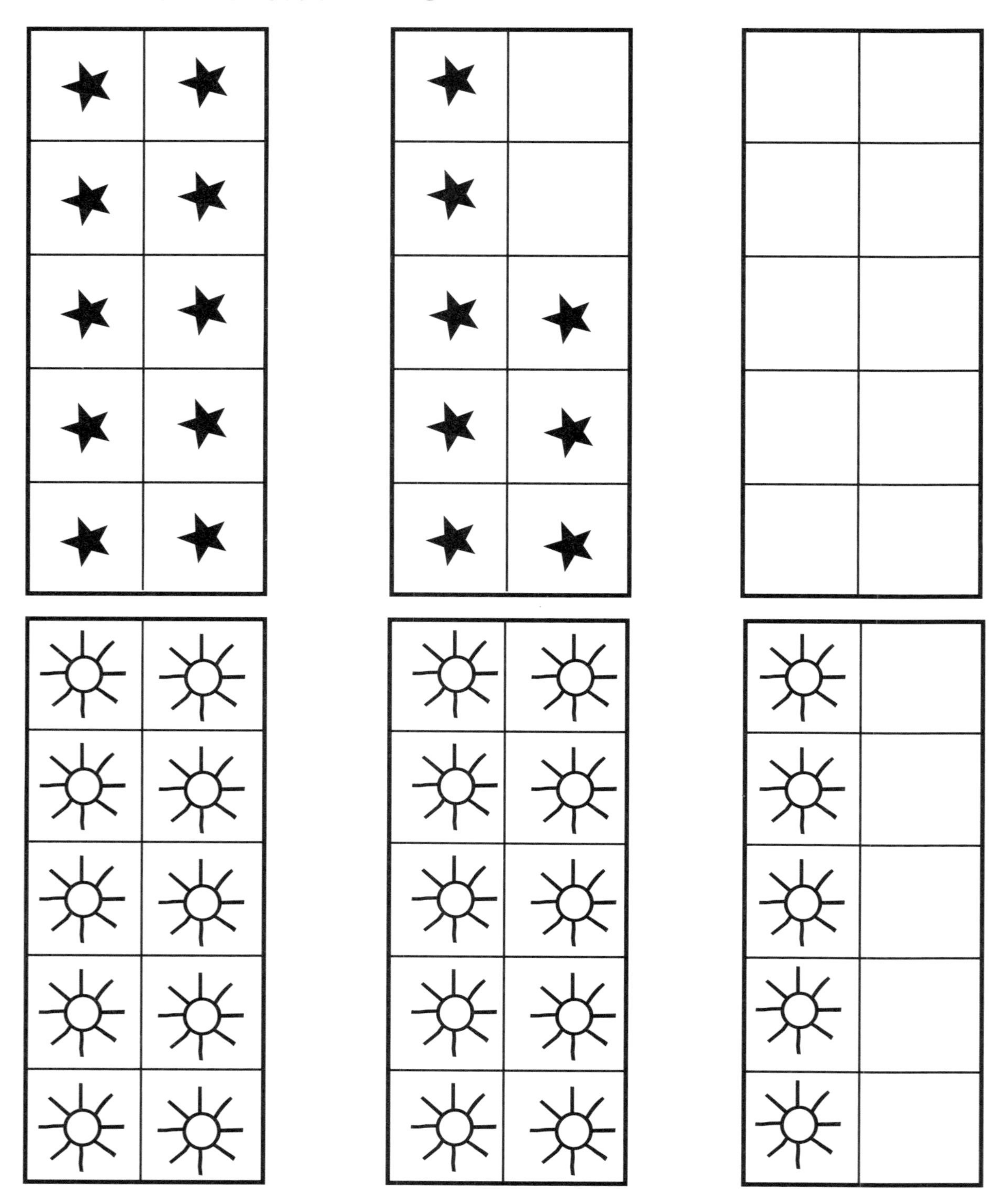

How many stars do you see? How many suns do you see?
How many squares are filled? How many are empty?
Can you make up a number sentence about the empty boxes?

Ten-frame Math 6

How many hearts do you see? How many umbrellas?
How many squares are filled? How many are empty?
Can you make up a number sentence about the hearts and the umbrellas?
Can you make up a number sentence about the empty boxes?

Pattern Block Shapes–Hexagons *Copy onto yellow cardstock.*

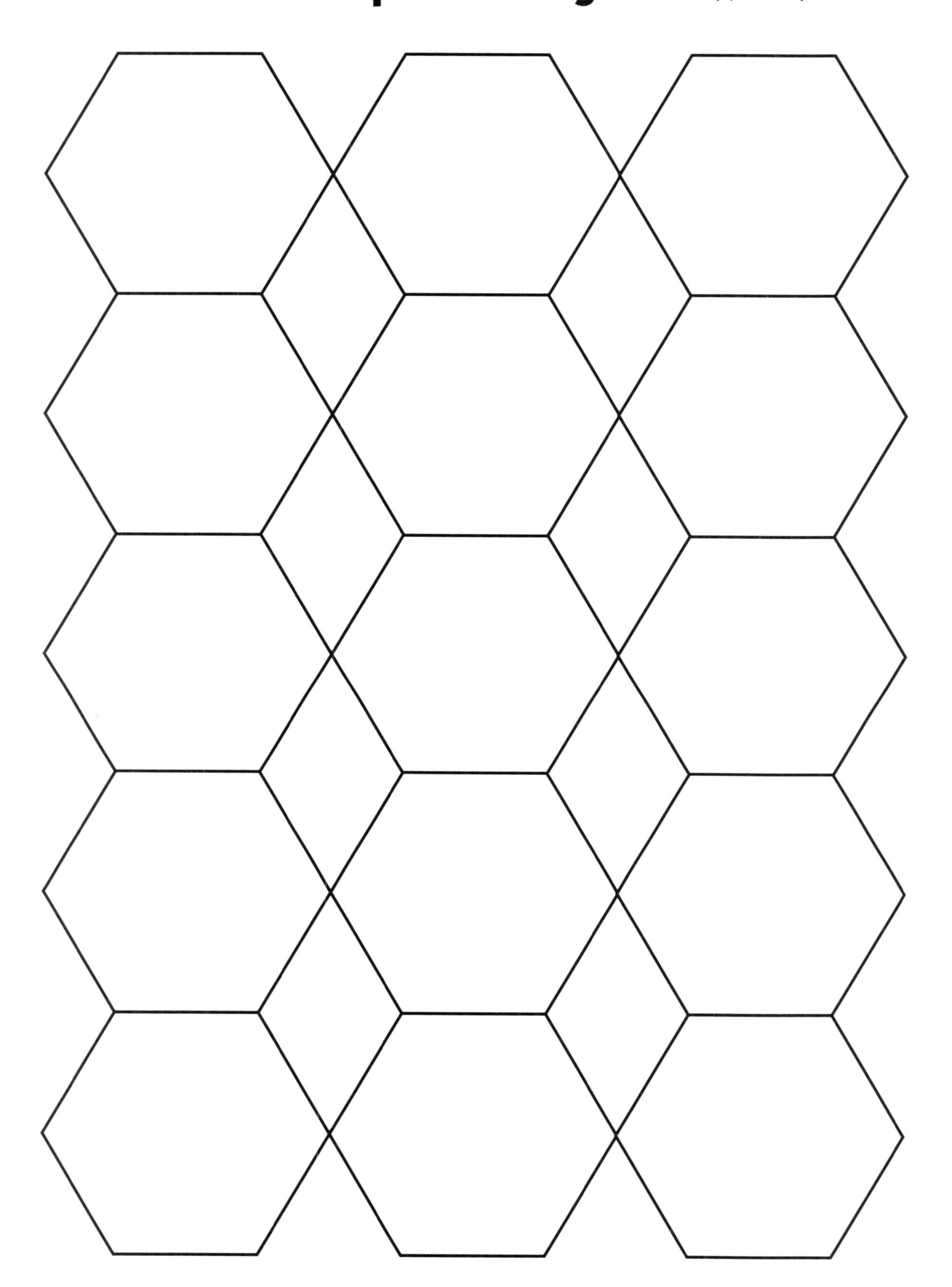

Pattern Block Shapes–Triangles

Copy onto green cardstock.

Pattern Block Shapes–Diamonds

Copy onto blue cardstock.

Pattern Block Shapes–Trapezoids

Copy onto red cardstock.

Pattern Block Shapes–Small Diamonds

Copy onto tan cardstock.

Large Block Paper

Bar Graph Extension

For students who need more space on a bar graph.
Cut and tape to the top of **Make a Graph**, S-10.

Bibliography

BOOKS FOR THE STUDENT

Bender, Lionel. ***Creatures of the Deep***. New York: Gloucester Press, 1989, 31pp.

Cave, Kathryn. ***Out for the Count***. New York: Simon and Schuster, 1991.

Carlstrom, Nancy White. ***Fish and Flamingo***. Boston: Little, Brown, 1993, unpaged.

Cazet, Denys. ***A Fish in His Pocket***. New York: Orchard Books, 1987, 32pp.

Ehlert, Lois. ***Fish Eyes: A Book You Can Count On***. San Diego: Harcourt Brace Jovanovich, 1990, 36pp.

Ferguson, Ava and Gregor Cailliet. ***Sharks and Rays of the Pacific Coast***. Monterey, CA: Monterey Bay Aquarium Natural History Series, 1990, 64pp.

Foster, Kelli C. ***Sometimes I Wish***. Hauppaughe, New York: Barron's, 1991, unpaged.

Gomi, Taro. ***Where's the Fish?*** New York: William Morrow, 1986, 26pp.

Hoban, Lillian, ***Arthur's Funny Money.*** New York: Harper Trophy, 1981.

Lionni, Leo. ***Swimmy.*** New York: Dragonfly Books™, 1963.

MacCarthy, Patricia. ***Ocean Parade.*** New York: Dial Books, 1990.

Pfister, Marcus. ***The Rainbow Fish***. New York: North-South Books, 1992, unpaged.

Silverstein, Shel. ***Where the Sidewalk Ends.*** New York: Snake Eye Music, Inc., 1974.

Sloat, Teri. ***From One to 100.*** Dutton: New York, 1991.

Turnage, Sheila. ***Trout the Magnificent***. San Diego: Harcourt Brace Jovanovich, 1984, 48pp.

Walton, Rick. ***How Many How Many How Many***. Cambridge: Candlewick Press, 1993.

BOOKS FOR THE TEACHER

Burns, Marilyn. ***Math and Literature (K-3).*** Math Solutions: Sausalito, California, 1992. This book shows how to connect mathematics with the imaginative ideas in picture books to develop problem-solving strategies with primary students. The first ten lessons include examples of student work. Another fourteen lesson outlines offer additional ideas to use with other literature titles.

Hinton, Jacki and Sue Rafferty. ***KinderCorner Math***™. Sundance: Littleton, MA, 1991. This book links literature with themes and seasons of the school year. Lesson ideas are based on 18 favorite literature titles, all readily available in paperback.

Kanter, Patsy and Jan Gillespie. ***Every Day Counts***™ ***Partner Games***. Great Source Education Group: Wilmington, MA, 1996. A collection of games that emphasize the development of number concepts.

McCabe, Jane and Christine Losq. ***Games for Number Sense.*** Great Source Education Group: Wilmington, MA, 1998. Develop number sense and addition and subtraction skills using adaptations of favorite games like dominoes and hopscotch. Includes ideas for literature-based problem solving.

Whitin, David and Sandra Wilde. ***Read Any Good Math Lately?*** Heinemann: Portsmouth, NH. 1992. If you have a well-stocked library, this volume will give you lots of ideas for new ways to use those wonderful old titles all the way through grade 6.

Whitin, David and Sandra Wilde. ***It's the Story That Counts.*** Heinemann: Portsmouth, NH, 1995. An updated companion volume to ***Read Any Good Math Lately?***, this resource provides an updated bibliography with more attention to picture books in print.

Index

SECTION 6

STUDENT ACTIVITY PAGES

(continued on next page)

Name ____________________ Date ____________________

Write About Addition and Subtraction

What do you know about adding?
Write about a time when you had to add.

What do you know about subtracting?
Write about a time when you had to subtract.

Name ____________________

Number Puzzle

1									
				25					
									50
				75					
									100

Name ____________________

Number Puzzle Pieces

Color the puzzle pieces

Then cut out the puzzle pieces.

Paste them where they belong in the Number Puzzle.

11	12	13	14
21	22	23	24

36	37	
	47	48

	53
62	63

98	99

66	67
76	77
86	87
96	97

	10
19	20

Name ____________________

Hundred Chart Puzzlers

Add our digits and you always get 10.
What numbers are we? Color us red.

Subtract our digits and you always get 2.
What numbers are we? Color us yellow.

1	2	3	4	5	6	7	8	9	10
11	12	13	14	15	16	17	18	19	20
21	22	23	24	25	26	27	28	29	30
31	32	33	34	35	36	37	38	39	40
41	42	43	44	45	46	47	48	49	50
51	52	53	54	55	56	57	58	59	60
61	62	63	64	65	66	67	68	69	70
71	72	73	74	75	76	77	78	79	80
81	82	83	84	85	86	87	88	89	90
91	92	93	94	95	96	97	98	99	100

Do you see a dragonfly?

Name ____________________

More Hundred Chart Puzzlers

Follow the directions.

Find each number.

Color each box green.

A. Start at 63.

B. Add 11.

C. Add 11 more.

D. Add 1.

E. Add 1.

F. Subtract 9.

G. Subtract 9.

H. Subtract 11.

I. Subtract 11 more.

J. Subtract 1.

K. Subtract 1 more.

L. Add 9.

M. Add 18 more.

N. Subtract 10.

O. Subtract 10 more.

1	2	3	4	5	6	7	8	9	10
11	12	13	14	15	16	17	18	19	20
21	22	23	24	25	26	27	28	29	30
31	32	33	34	35	36	37	38	39	40
41	42	43	44	45	46	47	48	49	50
51	52	53	54	55	56	57	58	59	60
61	62	63	64	65	66	67	68	69	70
71	72	73	74	75	76	77	78	79	80
81	82	83	84	85	86	87	88	89	90
91	92	93	94	95	96	97	98	99	100

What picture did you make?

____

Name ____________________

Keep the Difference

I played with ______________________________

I rolled	My partner rolled	Difference
$3 + 6 = 9$ (9 circled)	$6 + 6 = 12$ (12 circled)	$12 - 9 = 3$

Estimating Fish Area

Name ______________________

Make a New Ten

Name ____________________

I start with	I add	Now I have	Can I make a new ten?

Name ______________________

Big Fish and Little Fish

How many pieces are in your big fish?

Use the ten-frames to show how many.

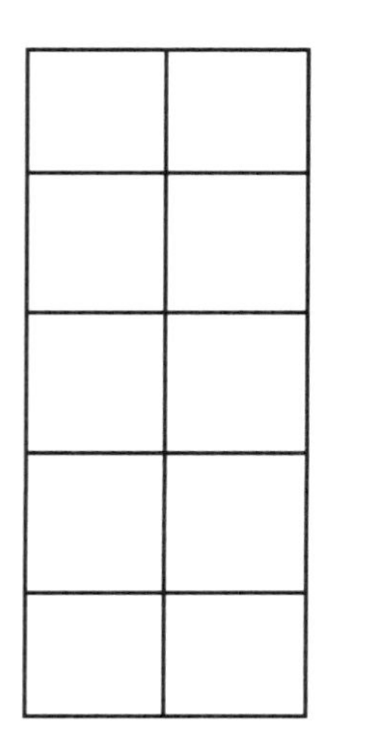
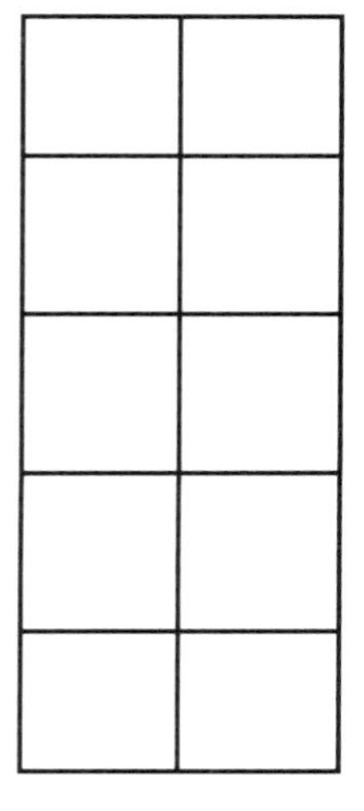

How many pieces are in your little fish?

Use the ten-frames to show how many.

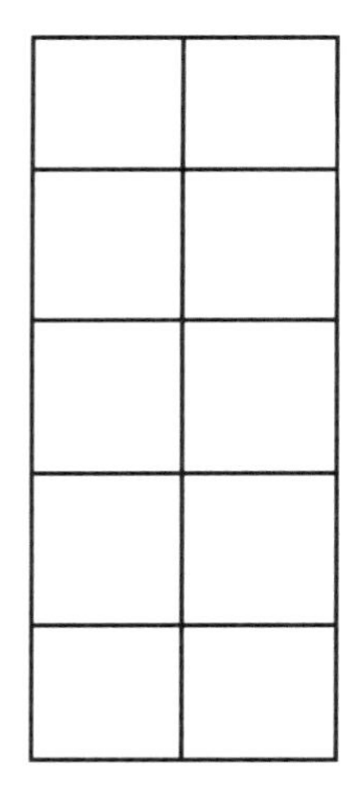
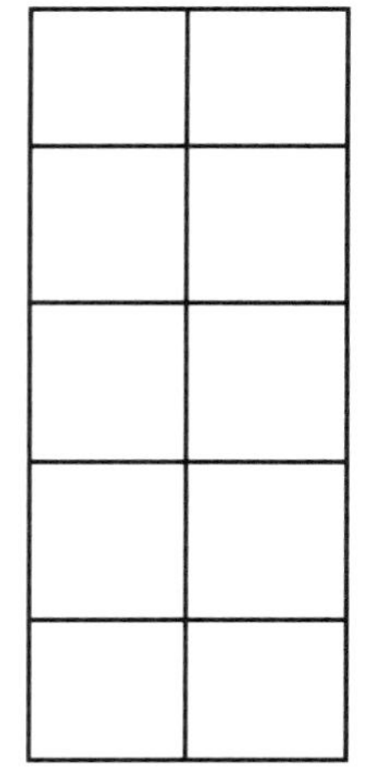

How many pattern pieces did you use to make your big and little fish?

Write a number sentence that tells how many pattern pieces in all.

____

Make a Graph

Name ______________________

Name ____________________ Date ____________________

What Does Your Graph Show?

Look at your fish graph.

How many of each shape did you use?

I used __________ green triangles.

I used __________ blue diamonds.

I used __________ yellow hexagons.

I used __________ red trapezoids.

I used __________ tan diamonds.

Write two sentences about shapes you used.

____

____

____

____

____

____

Name ____________________ Date ____________________

Tell About Your Graph

Use ten-frames and colored markers.

Show how many green triangles you used.

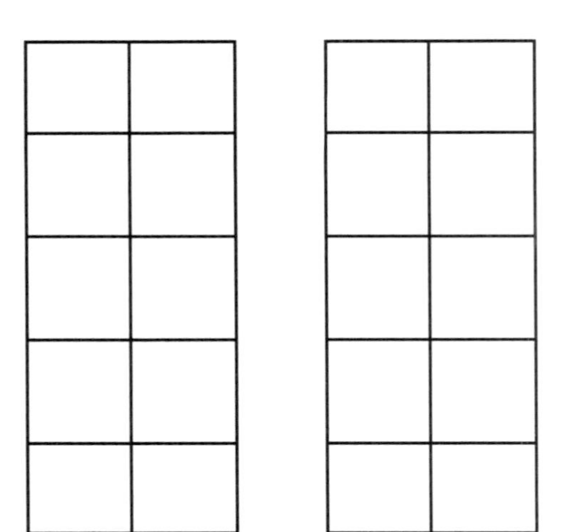

Show how many blue diamonds you used.

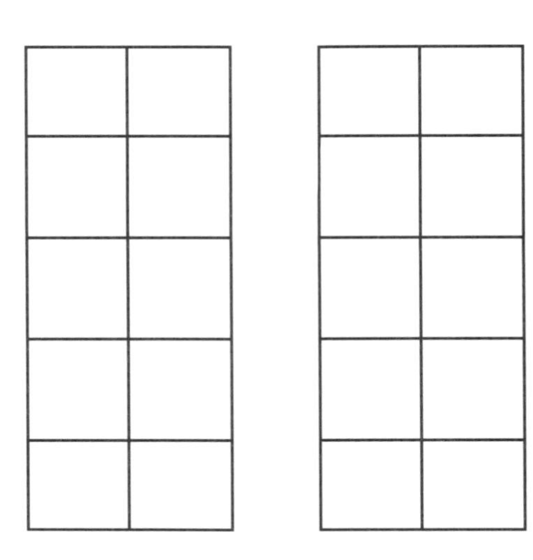

Did you use more green triangles or more blue diamonds? How many more? How do you know?

____

____

____

____

____

____

Name ____________________ Date ____________________

Tell More About Your Graph

Use ten-frames and colored markers.

Show how many yellow hexagons you used.

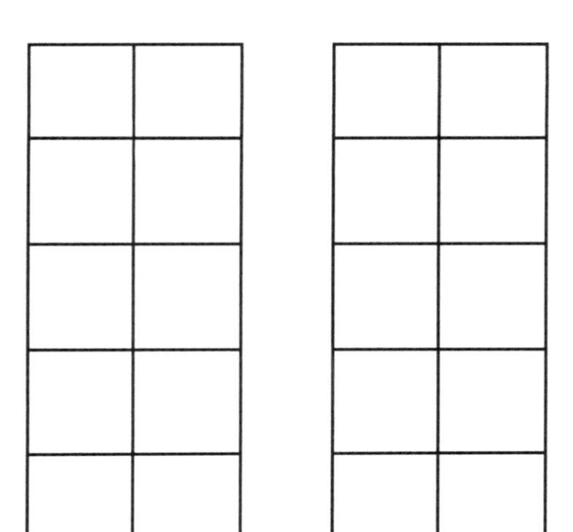

Show how many tan diamonds you used.

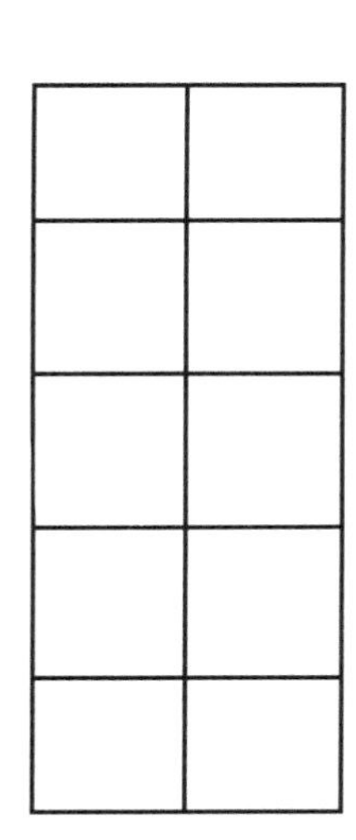

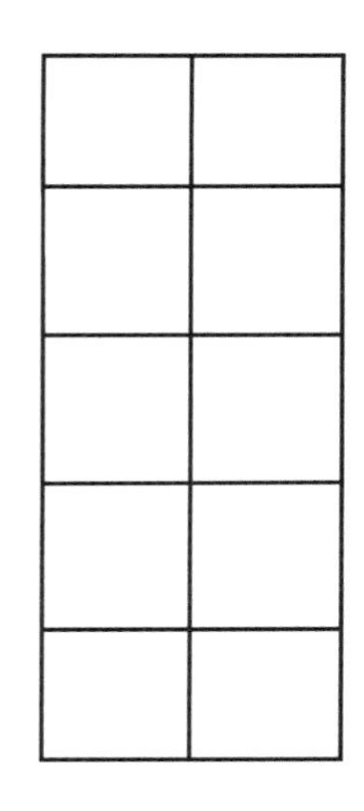

Did you use more yellow hexagons or tan diamonds?
How many more?

____

____

____

____

____

____

Ways to Make Ten

Name ____________________

Color to show tens.

You need ten pennies and crayons in two colors.

Use one color for heads and a different color for tails.

Use heads and tails on pennies to show tens.

How many ways can you find?

Color ten-frames to show the ways.

Then write a number sentence for each way.

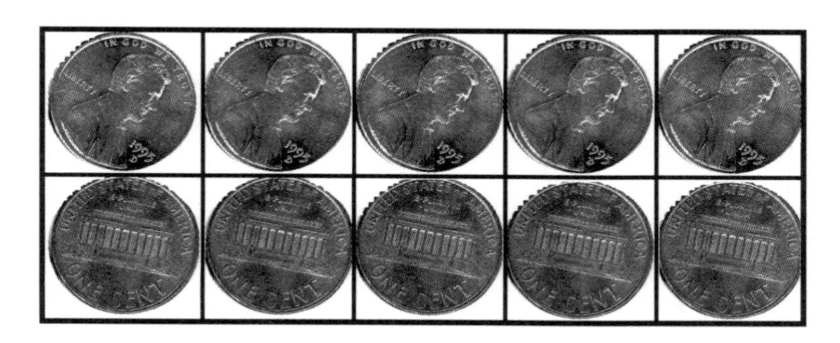

$$\begin{array}{rl} 5 & \text{heads} \\ +\ 5 & \text{tails} \\ \hline 10 & \text{pennies} \end{array}$$

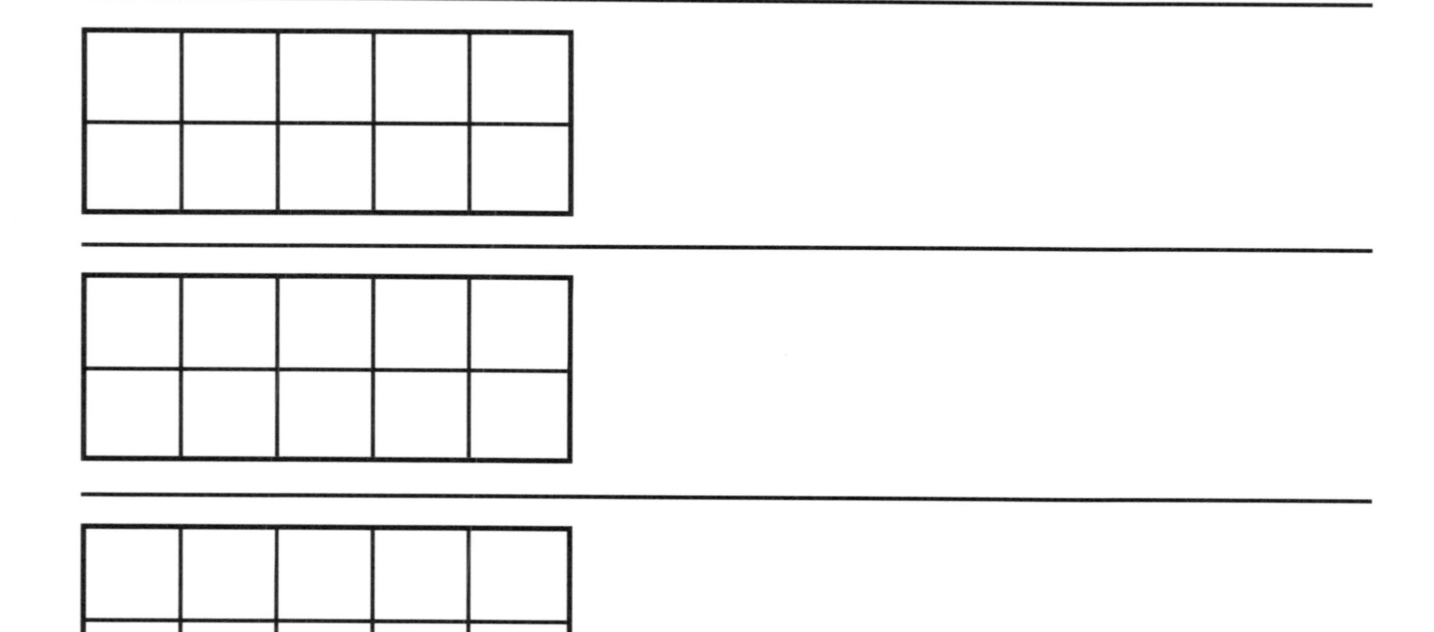

More Ways to Make Ten

Name ______________________

Show heads and tails.

Write a number sentence for each picture.

Name ____________________

Marshmallows and Macaroni

Use ten-frames.

Count your marshmallows.

Record what you see.

Write the number.

____

Use ten-frames.

Count your macaroni.

Record what you see.

Write the number.

____

How many marshmallows and macaroni do you have in all?

Write an equation in the box.

Do you have more macaroni or more marshmallows?

How many more?

Name ____________________

Writing Equations for Pictures

How many of each shape do you see?

Write two equations about circles and squares.

What do your equations say about circles and squares?

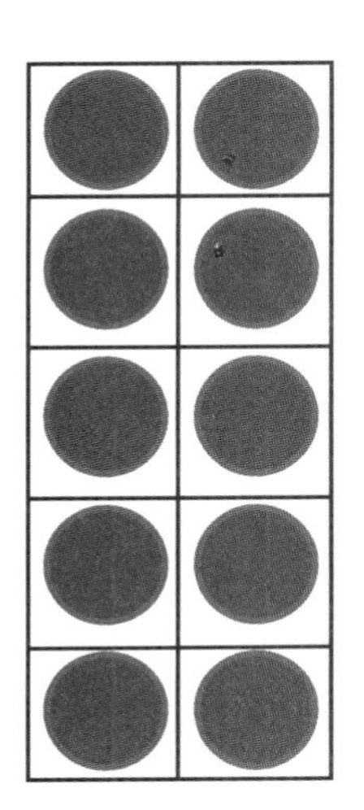
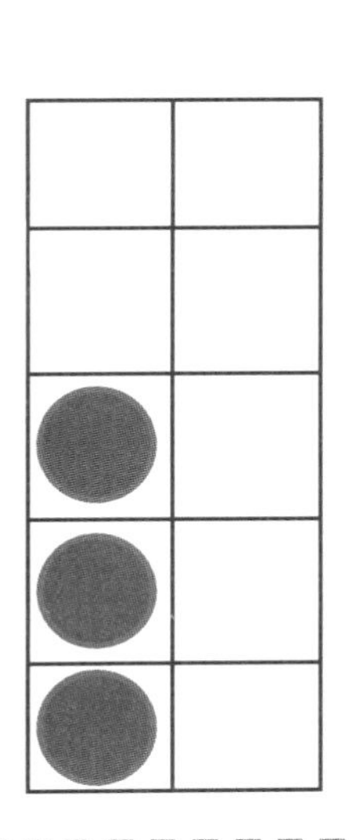
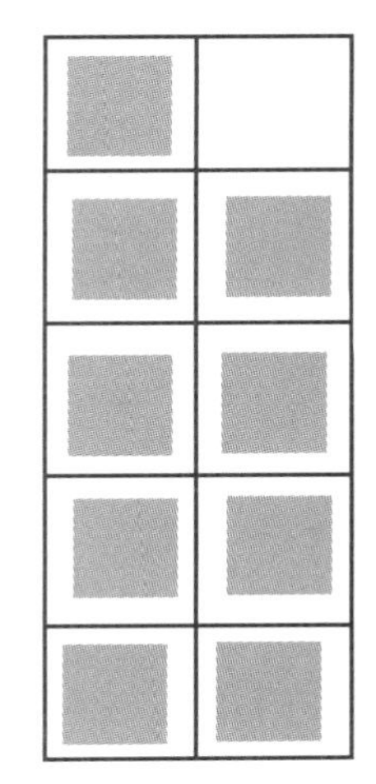

$13 - 9 = 4$

There are 4 more circles than squares.

$$\begin{array}{rl} 13 & \text{circles} \\ +\ 9 & \text{squares} \\ \hline 22 & \text{shapes in all} \end{array}$$

Now look at this picture.

Write two equations.

What do your equations say?

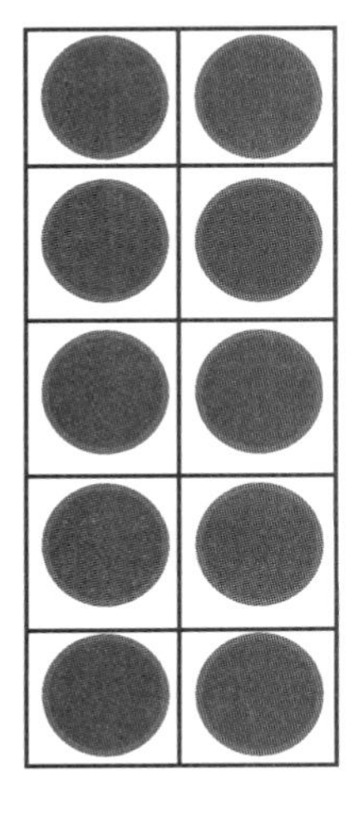
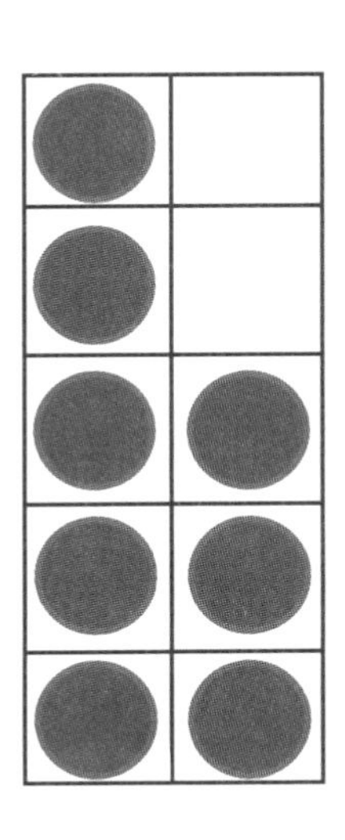
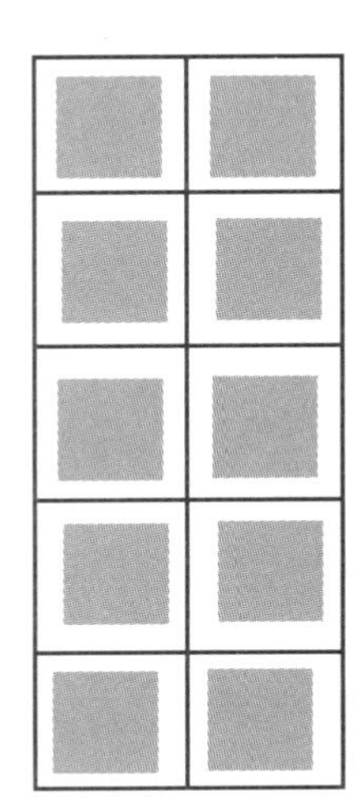
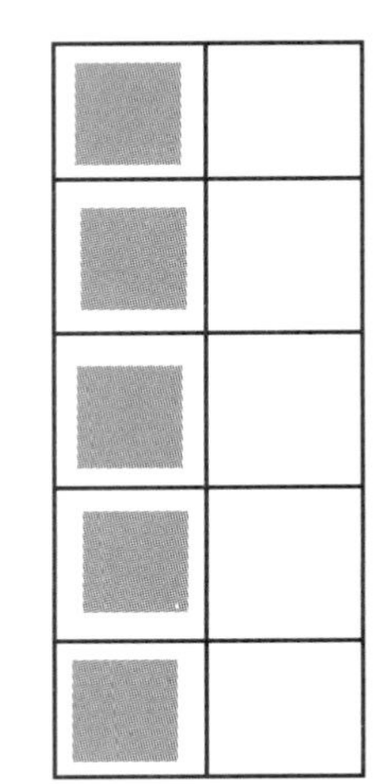

Name ____________________

Ten-frame Addition

You need a box of crayons.

Show each equation in ten-frames.

Then solve.

1. $\begin{array}{r} 14 \\ +\ 9 \\ \hline \end{array}$ red dots / blue dots

   14 red dots
   \+ 9 blue dots

2. 8 green dots
   \+ 25 brown dots

3. 21 orange dots
   \+ 18 brown dots

4. 9 orange dots
   \+ 17 purple dots

5. 23 black dots
   \+ 9 yellow dots

6. 27 red dots
   \+ 7 yellow dots

7. Write a story about 50 dots. Share your story with a partner. Solve each other's stories.

Name ______________________

Show Ten-frame Addition

1.

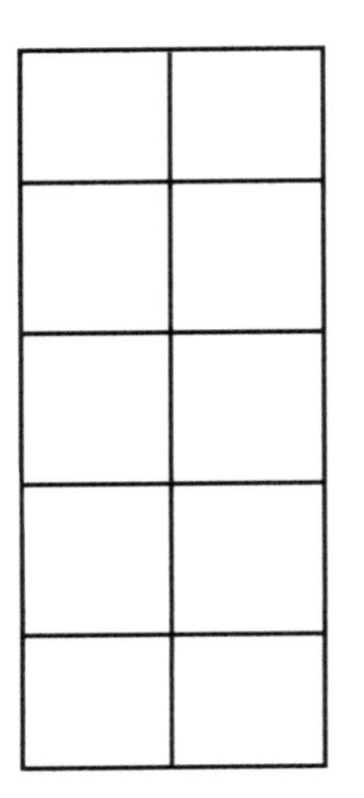

Write a number sentence for your picture.

2.

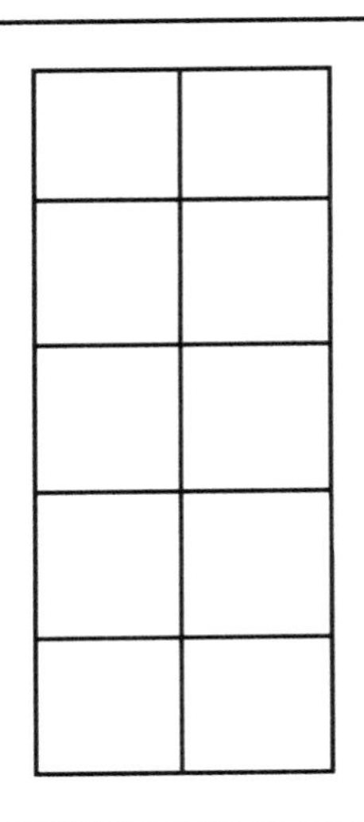

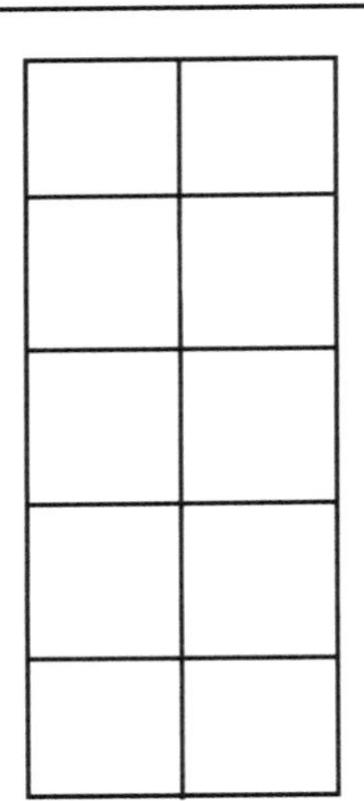

Write a number sentence for your picture.

3.

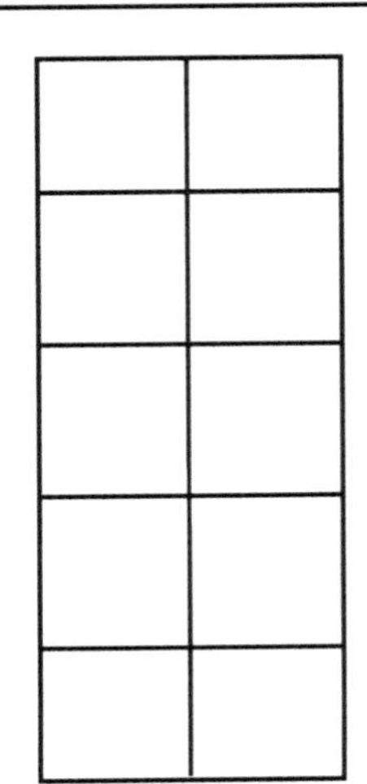

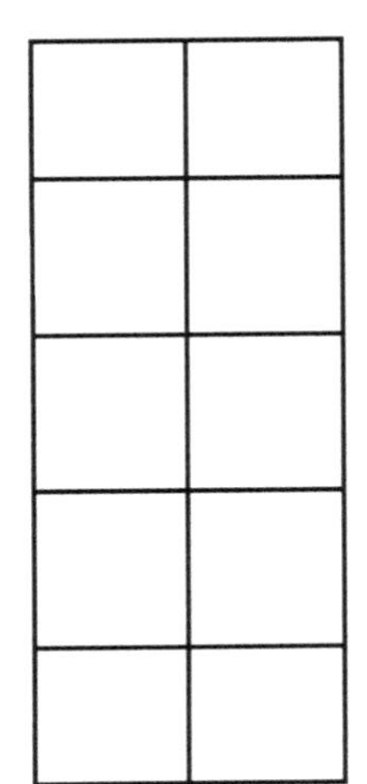

Write a number sentence for your picture.

Name ______________________

Show More Ten-frame Addition

4.

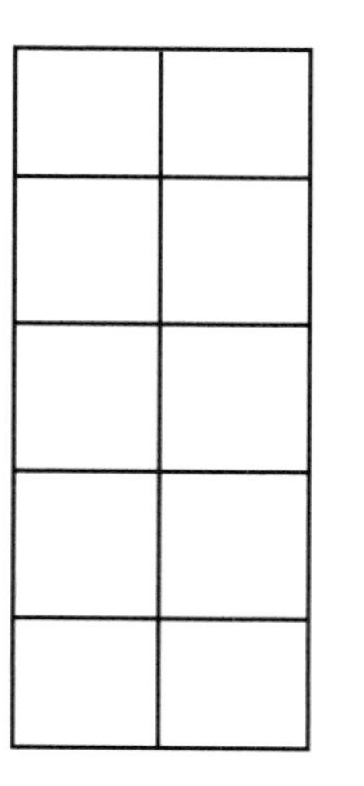

Write a number sentence for your picture.

5.

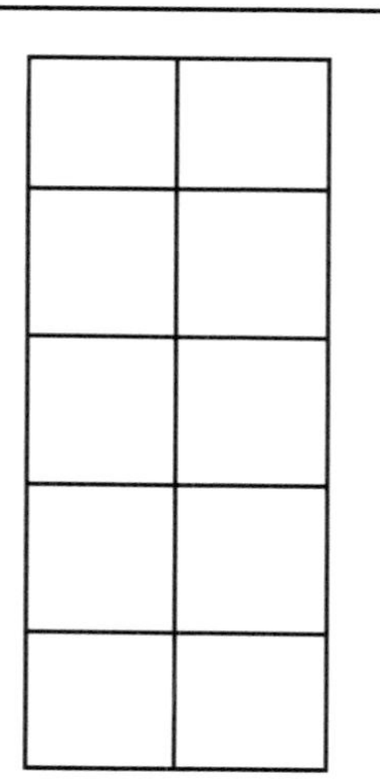

Write a number sentence for your picture.

6.

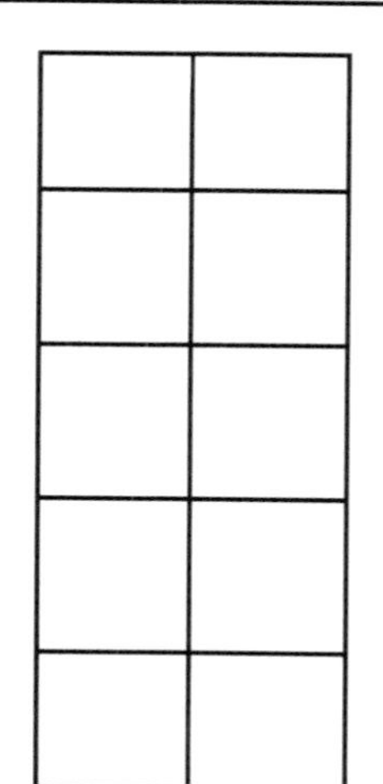

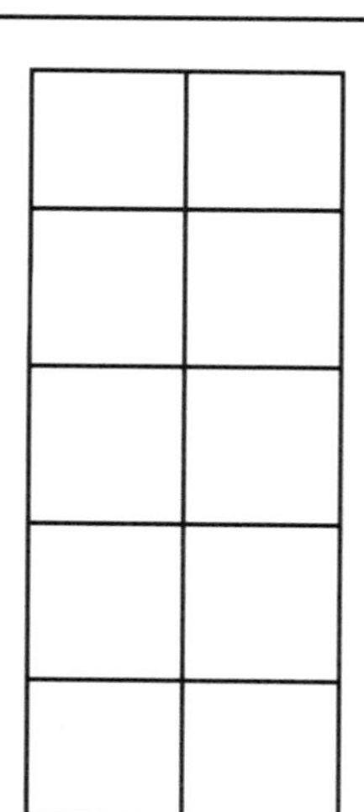

Write a number sentence for your picture.

Name ____________________

Marshmallow Take-Away

Use marshmallows and a number line or a yardstick.

Act out each story.

Write how many are left.

1. 12 marshmallows
   10 rolled away
   ______ are left

2. 14 marshmallows
   10 rolled away
   ______ are left

3. 16 marshmallows
   10 rolled away
   ______ are left

4. 18 marshmallows
   10 rolled away
   ______ are left

What patterns do you see in the marshmallow stories?

Make up your own marshmallow story to solve.

Name ____________________

More Marshmallow Take-Away

Use a number line and marshmallows.

Write an equation for each story.

Solve each story.

1. 17 marshmallows
   4 rolled away

2. 15 marshmallows
   9 rolled away

3. 19 marshmallows
   13 rolled away

4. 22 marshmallows
   18 rolled away

5. 31 marshmallows
   18 rolled away

6. 25 marshmallows
   6 rolled away

7. I wonder how many marshmallows rolled away in all the stories. How would you figure it out?

____

Name ____________________

Arthur Counts His Money

Read pages 1-15 in ***Arthur's Funny Money***.

Then answer these questions.

1. How much money did Arthur have in his piggy bank?

$ ____________________

2. Which coins might Arthur have in his bank?

Name ____________________

Arthur Sets Up for Business

Read pages 16-19 in ***Arthur's Funny Money***.

Then answer these questions.

1. What does Arthur do to set up his business?

2. How much money does he spend on supplies?

3. How much more money will he need to earn?

Name ____________________

Keep Track of Arthur's Money

Read pages 20-44 in ***Arthur's Funny Money***.

Record Arthur's earnings and expenses.

What he did	How much he earned	or How much he spent
bought soap		53¢
bought Brillo		27¢
washed bike and trike	42¢	

How much money did Arthur earn?

How did you figure it out?

____

Name ____________

Keep Track of Money

Finish reading ***Arthur's Funny Money***. Then answer these questions.

1. Arthur had $4.48. How much more money would he need to reach $5.00?

2. Arthur bought a cap and shirt for $4.25. How much less than $5.00 did the cap and shirt cost?

3. Suppose Arthur had 30¢ and licorice twists cost 5¢ each. How many licorice twists could he buy?

   Explain how you figured it out.

Name ______________________ Date ______________________

What Did I Learn?

My favorite activity was ____

I liked it best because

____

____

____

____

The hardest thing about adding and subtracting is

____

____

____

____

I want to learn more about

____

____

____

____

Name ____________________ Date ____________________

I Can Solve Story Problems

1. 14 shapes make an eel. 13 shapes make a sea star. How many shapes are there together? __________

2. Kate got 13 fish crackers. Then she got 9 more. How many fish crackers does Kate have? __________

3. Matt counted out 23 marshmallows. Then he ate 9. How many does he have left? __________

4. Arthur had 97¢. He spent 25¢ on licorice twists and 42¢ on soap. How much money does Arthur have left? __________

Name ______________________ Date ______________________

Ten-frame Math

1. How many pencils do you see?
2. How many flowers do you see?
3. How many more flowers than pencils do you see?

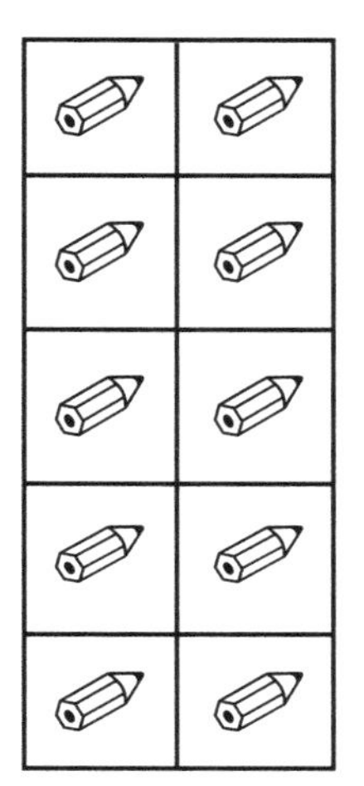

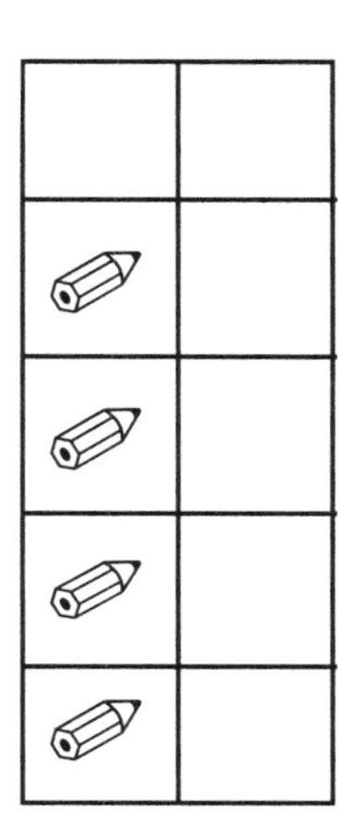

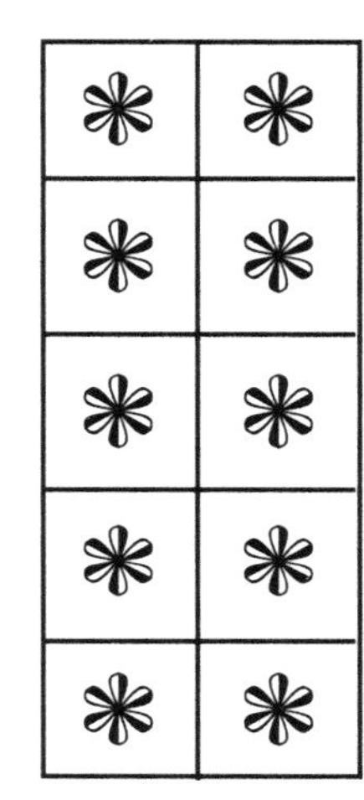

4. Write a number sentence about the pictures.
   Explain what it means.

Now solve these.

5. 19 red dots
   \+ 6 blue dots

6. 5 orange dots
   \+ 17 purple dots

7. 19 fish
   − 6 swam away

8. 25 fish
   − 17 swam away

Name ________________ Date ________________

Make a Graph

1. Use pattern pieces. Make fish like the ones in the picture.
2. Make a graph about the shapes you used.
3. Tell what your graph shows.

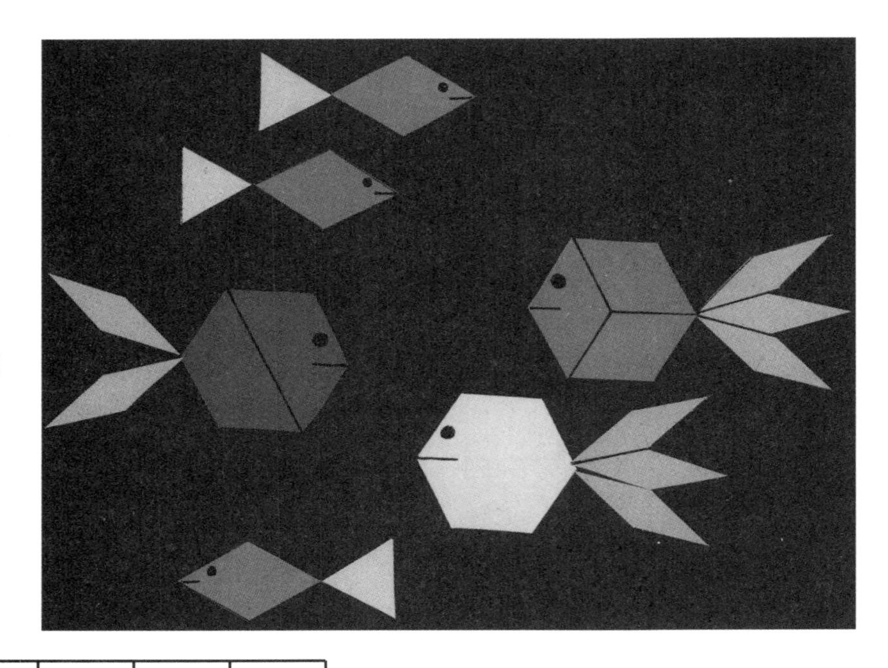

My graph shows

____

____

____

____

____

____

____